Using R as a GIS
A Beginner's Guide to
Mapping & Spatial Analysis

Dr. Nick Bearman

Credits & Copyright

Using R as a GIS

A Beginner's Guide to Mapping & Spatial Analysis

by Dr. Nick Bearman
Published by Locate Press

Direct permission requests to info@locatepress.com or mail:
Locate Press, B102 5212 - 48 ST. Suite 126
Red Deer, AB, Canada, T4N 7C3

Publisher Tyler Mitchell
Editor Keith Mitchell
Cover & Production Design Nathan Watson & Julie Springer
Interior Design Based on Memoir-LATEXdocument class
Publisher Website http://locatepress.com
Book Website http://locatepress.com/book/rgis

Version: 1be836a (2024-11-16)

Contents

Introduction

Welcome to Using R as a GIS!

This book is all about the scripting language R, and how you can use it as a Geographic Information System (GIS) to manage, display, symbolise, and analyse spatial data. While initially starting out as a statistics package, now R can also create maps and perform spatial analysis - much like the tools you would find in QGIS or ArcGIS Pro.

In this book, we will cover the basics of GIS and R, as well as some more advanced visualisation and analysis techniques. We assume no prior knowledge of GIS or scripting. This book is also suitable if you have previously used a graphical desktop GIS, but want to move to a script-based GIS. Two of the key benefits for adopting a script-based approach is: code reusability, and sharing of knowledge with others, e.g. sharing code from a research project.

This book is split into five parts:

- **Part 1: What is GIS & Spatial Data?** Covers some of the key pieces of theory you need to know to work with a GIS successfully. Chapter 1 gives an overview of the book and introduces some basic, core techniques in R. You will also start creating maps in Chapter 3, and customising them so they show what you want them to.
- **Part 2: Spatial Data** Covers building more advanced maps, and making use of raster and vector data. It also covers creating interactive maps in R and the differences between the SF and SP libraries.
- **Part 3: Advanced Concepts** Applies some of the GIS techniques we have learned to more advanced programming concepts in R, including loops (repeating code multiple times with different data sets), if/else functions (running different bits of code depending on the data), and spatial data wrangling (getting your data into a usable format in R).

- **Part 4: Project Management** Covers some really important file management techniques, projects, version control, and R markdown. This will help you develop from running a series of scripts to do your analysis, to making full use of R and RStudio tools in your overall workflow.
- **Part 5: Next Steps** Covers some key hints and tips on where to find spatial data when you are working on your own project, what your next steps are after finishing the book, and some really useful hints and tips for finding information and help on the web.

Each chapter in these parts covers a specific topic, and many chapters have questions or an exercise at the end. It is highly recommended to do the exercises - the more practice you can do, the better! You can check your answers against the answer script available on the website.

This book is designed for people who are new to R, or even new to scripting and programming. No prior knowledge is needed! Ideally, you will know a little bit of GIS already, and maybe have opened QGIS or ArcGIS. However don't worry if you haven't, we will cover the key GIS terminology that you need to know.

You can work through this book chapter by chapter, or simply use it as a reference. Use the book in whatever way works for you!

Part 1

What is GIS & Spatial Data?

1. What is R?

1.1 Learning Outcomes

After reading this chapter, you will:

- Understand what R is and how it works.
- Be aware of how R enables reproducible research.
- Understand how to use R & RStudio.
- Know some R basics.

In this chapter, we'll discuss what R is and why it's different from many other Geographic Information System (GIS) programs.[1] While R has some limitations, it also has many benefits and advantages compared to other GIS programs. We will discuss some of these advantages to set the scene for the rest of the book.

We will also cover some of the basic concepts of using R and provide links to further resources if you want to learn more foundational concepts in R before moving on to visualizing spatial data using R in Chapter 3.

For this sample chapter, we will also include some demo code showing you how to make maps using R.

1.2 R the Software

The R programming language works in a fundamentally different way to many other GIS-related programs you may have used. Rather than having a Graphical User Interface (GUI) with menus, buttons, and wizards to do things, R uses a command line interface. You write a command, press Enter, and then R does something.

---

[1]The R Project for Statistical Computing: r-project.org

This might be quite different from what you are used to, but don't be put off by this! If you're new to scripting or programming, it takes a bit of getting used to, but I hope I can show you the many benefits this approach offers over other GIS programs. If you've done any scripting or programming before, you will find that experience is very useful in learning to use R.

The command line approach does have some disadvantages. One of the things I found the hardest when I started to learn R was remembering the different commands. There is a glossary at the end of this book to help with this where all the R commands are listed, with examples of how they work.

Commands in R are also case sensitive, which means that the read_st() function is different from the read_ST() function!

However, while there are some disadvantages with a command line approach, there are also many advantages it offers over other GIS programs. R is very easily scriptable, which means that because you use written commands to do your analysis, it is very easy to track exactly what you did, and then share this code with others. For example, if your colleague needs to do the same analysis that you did, you can send them the script and it will repeat the commands exactly how you performed them.

You can also share code with your future self. By this, I mean that you can repeat the same analysis you did 6 months ago, without needing to remember exactly what you did or needing to make copious notes to reproduce the steps. It is all there in the script.

Finally, R also allows you to use computer programming concepts like loops. Loops allow you to repeat your code with different data. For example, if you have a table of 20 variables for the same geographic area, you can create a map in R for one of those variables. You can then use a loop to run that code 20 times to create 20 different maps, one for each variable. It would be a lot more work if you wanted to do that in QGIS or ArcGIS. You'll be learning how to use loops to create multiple maps within this book - see Chapter 8 for details.

1.3 Reproducible Research

Reproducible Research is a term that has become increasingly popular in academic circles since 2010, with the term first being used in 1992 [4], and the concept referred to as early as 1968 [10]. It has recently come to the forefront in GIScience with the developments of data science and advanced computational techniques [3]. However, the concept of reproducible research is key to all academic research. It has been around for a long time, underpinning scientific research from the beginning of its history.

The concept of reproducible research is that anyone reading an academic paper, with a suitable level of knowledge, should be able to recreate that research. Usually this refers to experimentation, but now also refers to software and statistical analysis.

R is great at reproducible research because the script records everything you do. By running the script, you can reproduce the data and results. Ensuring that your research is reproducible shows that it is good science. Some conferences and journals even ask you to submit your code and data along with your article, so that the reviewers can reproduce your results. R is a great language to do this in - although the principle works equally well with any programming language.

QGIS: Reproducible Research

If you're moving to R from QGIS, you might wonder whether QGIS can be used to create reproducible research. My answer is, "yes and no." It is possible to use QGIS for reproducible research, but making your work truly reproducible is quite tricky. QGIS is open source software, which is great for reproducible research. Anyone can download and install it to replicate your work, so this is not a problem.

Some may argue that access to technology could be an issue here, and if people can't access computer technology (such as a laptop or internet connection), then the work is not really reproducible. I would say a standard laptop capable of running QGIS is not too much of a requirement, and we have to draw the line somewhere. Of course, it may be slightly different considering resource intensive analysis like high-performance computing. However, this is beyond the scope of this book!

The other element is the work itself; how easy is it to replicate? If you have a script then this process is straightforward. You run the script, which repeats exactly what you did by repeating the same commands, and you get the same result at the end. QGIS is not primarily a scripting program, it uses a graphical user interface. It is challenging to record exactly what you did, and for another user to replicate it. You can do scripting (using Python) within QGIS, but it is not the core focus of the product. So in my view, you can't create truly reproducible work in QGIS. QGIS does have many other advantages, but this is a book about R. See Chapter 7: GIS software in Bearman (2021) [1] for a comparison of different GIS software.

> **ArcGIS: Reproducible Research**
>
> If you're moving to R from ArcGIS, you might wonder whether ArcGIS can be used to create reproducible research. Fundamentally, I would say no. A key element of reproducible research is open source software, which ArcGIS is not. ArcGIS is a commercial piece of software, which is often available to academics at a heavily discounted price.
>
> For a piece of research to be reproducible, someone needs to have the knowledge, data, process, and software to replicate your results. ArcGIS is primarily a graphical interface. This makes scripting your work more difficult, although not impossible. You can use Python, but this is trickier to do - see also the box on the previous page about QGIS.
>
> More importantly, if a user doesn't have access to the software you used, they cannot replicate your work. Since ArcGIS is not an open source product they have to buy the software to use it. If they do not have access to ArcGIS, then they can't reproduce the research. Therefore, I would say that ArcGIS (or any closed source software) cannot be used in truly reproducible work.

Another key aspect of reproducible research is version control. This allows multiple people to contribute to a project, and also allows you to 'snapshot' a specific version of your code so that you can always go back to it. See Chapter 12: Version Control for more information.

1.4 What is R & RStudio?

R has been around in some form since 1993[2]. It was originally designed as a statistics programming language, allowing users to perform all kinds of statistical analysis through a command line interface. One of the key features of R is the flexibility it offers through libraries that allow users to develop custom commands. Writing these libraries is not much more difficult than writing R code. Many users have written libraries for R, with 18,410 available at the time of writing[3]. This means that R can be used for many types of statistical analysis, including analysis for very niche applications.

The R system also allows you to pick and choose which libraries to load

---

[2]R programming language: wikipedia.org/wiki/R_(programming_language)

[3]18,410 libraries listed at cran.r-project.org/web/packages (May 2022).

at any one time. This ensures that R remains relatively quick to run and doesn't become overloaded with all the additional code provided by libraries. Particularly relevant to us, some of these libraries enable R to handle spatial data and create maps. We will explore this further in Chapter 3.

For the moment, just know that you can load libraries into R. These libraries provide R with additional functionality, including handling, mapping, and analyzing GIS data.

Since 2010, any publication discussing R includes a discussion of *RStudio*. R and RStudio are two separate programs but they're closely interrelated, so much so that when people talk about 'using R', it is commonly implied that they're also using RStudio.

R is the actual programming language itself, and the program named R understands the code we write and runs it. RStudio is an Integrated Development Environment (IDE) that acts as an interface running on top of R. It provides a nice environment for R scripts that makes it easier to write and work with them.

If you do any programming in other languages, you will find that they often have an IDE as well. A single programming language can have many different IDEs. Python has several IDEs, for example. There are several IDEs for R; however, RStudio is by far the most popular. Figure 1.1, on the facing page shows the differences between the R and RStudio interfaces.

RStudio has many more options, tools, and windows than R, which is quite bare bones by comparison. We will be working with RStudio when we do the exercises in this book, but it will be R that is actually doing the processing. RStudio is just the interface to make it easier for us to work with R.

1.5   Using R & RStudio

This next section will contain some code snippets as well as bits of text. Feel free to use this in whatever way works best for you. You can read through the text and try out the code in RStudio as you go, or read through all the text first and then try out the code. The code is also available in `ch1-what-is-r-script.R`.

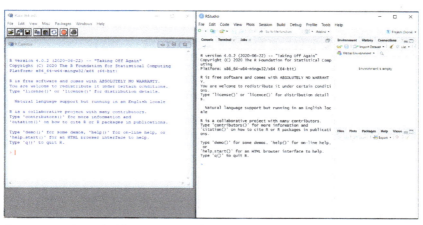

Figure 1.1: The R (left) and RStudio (right) interfaces. When people talk about working in R, most of the time they are using RStudio to write and run their code.

When you first open RStudio there are three key sections: *Console*, *Environment*, and *Files*. See their locations in figure 1.2, on the next page.

The *Console* is where you type in your commands, and R produces feedback. For example, if you type:

3+4

in the *Console* and press enter, R will respond:

[1] 7

While you can type individual commands in R, you can also collect several commands together in a script. We will expand on scripting in Chapter 3.

The *Environment* (top-right, figure 1.2, on the following page) lists the variables R is currently using. R uses the concept of variables to store data. This could be anything from a single number, to a series of numbers, to words, to a whole spatial data layer. The same concept is used to set up or read in variables via the <- "assign to" symbol.

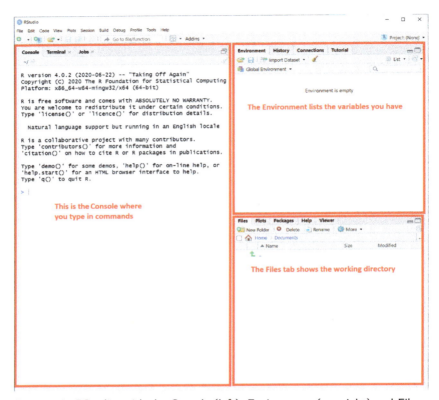

Figure 1.2: RStudio, with the Console (left), Environment (top-right) and Files (bottom-right) tabs labelled

The *Files* tab (bottom-right, figure 1.2) shows the working directory. The working directory is a key concept in R - it is where R loads and saves files. You might have a different working directory for each project or topic you are working on. You can choose what the working directory is. Directory and folder mean the same thing, and those terms are used interchangeably in R.[4] When you load data into R, it will only look for files in the working directory. R also only saves newly created or modified files into the working directory.

---

[4]The term "directory" was used exclusively prior to 1995. When Microsoft released Windows 95, in 1995, they introduced the term "folder". The two have been used interchangeably ever since.

1.5.1 Installing R & RStudio

We'll need to install R and RStudio before moving further on in this book. I have instructions for installing R and RStudio available on my website[5]. If you have issues installing R, RStudio, or the sf and tmap libraries, you can use the website *Posit.Cloud* instead. Posit.Cloud is a website that allows you to run RStudio from within a web browser. Details on how to use Posit.Cloud are provided in the instructions listed in the link.

1.5.2 R Basics Worksheet

This worksheet takes you though some basic concepts and commands in R, with no prior knowledge of R needed to complete the work. However, if you want some more information on R, how it works, and basic commands, check out my Getting Started with R guide[6].

Type what you see below into the *Console* section of RStudio.

Any line prefixed with a # is a comment - a note for you, something the computer ignores. You don't need to type in these lines. However, commenting your code is really important. We will talk more about this when we get to scripts.

R can initially be used as a calculator. Enter the following into the left-hand side of the window, the section labelled *Console*:

```
6 + 8
```

You will see R outputs this:

```
[1] 14
```

Don't worry about the [1] for the moment, for now note that it precedes the result returned from the command we ran. R printed out 14 since this is the answer to the sum you typed in.

---

[5]Installing R and RStudio: nickbearman.github.io/installing-software/r-rstudio
[6]Getting Started with R: rpubs.com/nickbearman/gettingstartedwithr

We can also do multiplication:

```
5 * 4
[1] 20
```

Also note that `*` is the symbol for multiplication here. The last command asked R to perform the calculation '5 multiplied by 4'. Some other symbols include - for subtraction and / for division:

```
12 - 14
[1] -2

6 / 17
[1] 0.3529412
```

You can also assign the answers of the calculations to variables, and use those variables in calculations:

```
price <- 300
```

Here, the value 300 is stored in a variable called `price`. The `<-` symbol tells R: put the value on the right into the variable named on the left. It is typed with a < (less-than sign) immediately followed by a - (dash or minus). The variables are shown in the window labelled *Environment*, in the top right (see figure 1.3).

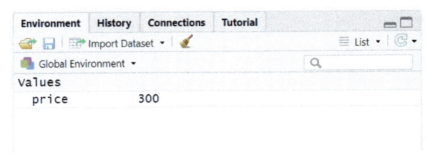

Figure 1.3: The price variable, listed in the Environment

We can use variables in later calculations. For example, to apply a 20% discount to this price you could enter the following:

```
price - price * 0.2
[1] 240
```

Or we can use intermediate variables to store the amount of the discount in a discount variable, and use that to calculate the final discounted price:

```
discount <- price * 0.2
price - discount
[1] 240
```

Along with individual numbers, R can also work with lists of numbers. Lists are specified using the c function. Suppose you have a list of house prices specified in thousands of pounds. You could store them in a variable called house.prices like this:

```
house.prices <- c(120,150,212,99,199,299,159)
```

Running this code will then show us what was stored in the house.prices variable:

```
house.prices
[1] 120 150 212  99 199 299 159
```

Note that there is no problem with full stops or a period in the middle of variable names.

You can then apply functions to the list of numbers we created:

```
mean(house.prices)
[1] 176.8571
```

If the house prices are in thousands of pounds, then this tells us that the mean house price is £176,900 (pounds).

1.5.3 The Data Frame

R has a way of storing related data in an object called a *data frame*. This is rather like a spreadsheet, which stores data together as a set of rows and columns.

Below is a CSV file of house prices and burglary rates, which you can
load into R. You can use a function called read.csv() which, as you
might guess, reads CSV files. Run the line of code below to load the
CSV file into a variable called hp.data.:

```
hp.data <- read.csv("https://locatepress.com/files/rgis/hpdata.csv")
```

It is always a good idea to check that data you've loaded came in okay.
To do this, you can preview the data set. The head command shows the
first six rows of the data:

```
head(hp.data)
```

```
  ID Burglary Price
1 21        0   200
2 24        7   130
3 31        0   200
4 32        0   200
5 78        6   180
6 80       19   140
```

You can click on the hp.data variable listed in the *Environment* window,
which will show the data in a new tab. You can also run the line of code
below to open a new tab in RStudio showing the data (figure 1.4, on
the next page):

```
View(hp.data)
```

You can also get a description of each column in the data set by using
the summary function:

```
summary(hp.data)
```

```
      ID            Burglary          Price
Min.   : 21.0   Min.   : 0.000   Min.   : 65.0
1st Qu.:615.5   1st Qu.: 0.000   1st Qu.:152.5
Median :846.5   Median : 0.000   Median :185.0
Mean   :654.0   Mean   : 5.644   Mean   :179.0
3rd Qu.:875.8   3rd Qu.: 7.000   3rd Qu.:210.0
Max.   :905.0   Max.   :37.000   Max.   :260.0
```

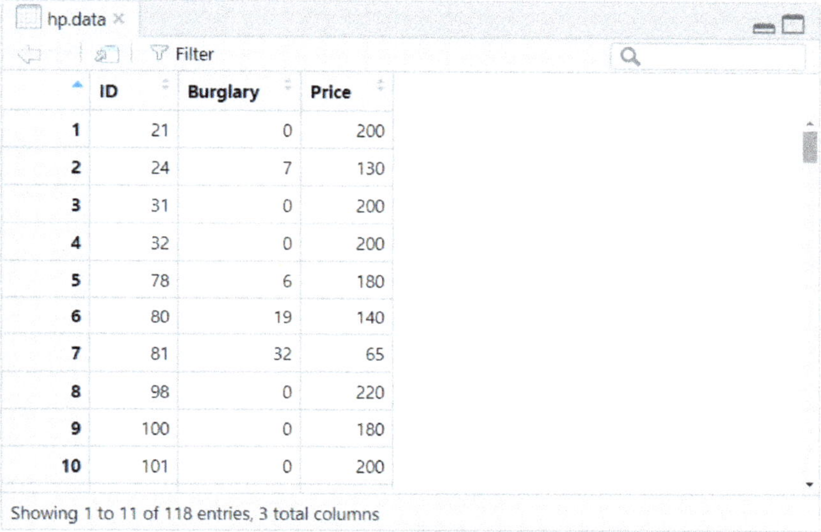

	hp.data ×			

	Filter		Q

	ID	Burglary	Price
1	21	0	200
2	24	7	130
3	31	0	200
4	32	0	200
5	78	6	180
6	80	19	140
7	81	32	65
8	98	0	220
9	100	0	180
10	101	0	200

Showing 1 to 11 of 118 entries, 3 total columns

Figure 1.4: The hp.data variable, shown as a table, using the View() function

For each column, a number of values are listed:

```
=======   ================================
Item      Description
=======   ================================
Min.      The smallest value in the column
1st. Qu.  The first quartile
Median    The median
Mean      The average of the column
3rd. Qu.  The third quartile
Max.      The largest value in the column
```

Based on these numbers, an impression of the spread of values of each variable can be obtained. In particular, it is possible to see that the median house price in St. Helens by neighbourhood ranges from £65,000 to £260,000. Half of the prices lie between £152,500 and £210,000. Also, it can be seen that since the median measured burglary rate is zero, then at least half of the areas had no burglaries in the month the counts were compiled.

Square brackets can be used to look at specific sections of the data frame, for example hp.data[1,] shows the first row and hp.data[,1] shows the first column. You can also delete and create new columns using the code below. Remember to use the head() command as we did earlier to look at the data frame. Try running the code below, and see what it does. The output has not been included here for brevity

```
# Create a new column in hp.data called "counciltax",
# storing the value NA.
hp.data$counciltax <- NA

# See what has happened after the change.
head(hp.data)

# Delete the "counciltax" column from hp.data.
hp.data$counciltax <- NULL

# See what has happened after the change.
head(hp.data)

# Rename the 3rd column in the dataframe to "Price-thousands"
colnames(hp.data)[3] <- "Price-thousands"

# See what has happened after the change.
head(hp.data)
```

1.6 Summary

In this chapter we have covered:

- An overview of how R works.
- What reproducible research is, and how it applies to GIS analysis in R.
- What RStudio is and how it relates to R.
- Basics for working with R commands.

There is much more we can do in R, but these are the building blocks you will base the rest of your learning on. Go on to Chapter 2 if you want to learn more about GIS, or skip to Chapter 3 if you want to jump right in with using R as a GIS!

2. What is GIS and Spatial Data?

2.1 Learning Outcomes

After reading this chapter, you will:

- Understand how we use GIS to think about space and location.
- Understand the importance of projections and coordinate systems within a GIS.
- Be aware of the different conceptual models that underpin vector and raster data.
- Understand the different advantages and disadvantages of vector and raster data.
- Be aware of different GIS-related file formats, such as shapefile and geopackage.
- Be aware of different data types, such as string and numeric data types, and understand how they can impact what we can do with spatial data.
- Understand the concept of scale and why it is important in GIS.

2.2 Introduction

This chapter covers some more theoretical and technical aspects of GIS: how it works, what we can use it for, map projections and coordinate systems, vector and raster data types, and spatial file types. I keep the GIS theory to a minimum, but this knowledge is really useful if you want to understand what is going on behind the scenes in a GIS and know how to fix problems when they happen.

We also cover the key concepts that you need to know before working with spatial data in R. We won't go into deep detail on these key concepts, but there are references if you would like further information.

A set of questions at the end of this chapter will test your knowledge, before implementing what you've learned in R in the next chapter.

2.3 How do we Think About Space?

There are two key elements to consider when thinking about space - how to specify a location (coordinates), and then how to conceptually think about objects in space (object-based or field-based conceptual models).

2.3.1 Projections

We use numbers, or *coordinates*, to represent locations. But first, we need to think about how we represent the world as a map. The Earth we live on is a 3D sphere, more-or-less[7]. However, most maps are flat pieces of paper. So how do we change the 3D spherical earth into a flat piece of paper? The process which we use to do this is called a *projection*, and there are many types of projections. Figure 2.1, on the facing page shows some common projections. You may be familiar with some names, and some maps probably look familiar.

The key thing to remember is that when dealing with global maps, no matter the projection we use, there will be some sort of distortion. It is just not possible to show a 3D sphere on a flat piece of paper without any distortions. Usually, either the area or the shape (e.g. of countries) can be maintained - but not both.

The *Mercator projection* (figure 2.1, on the next page) is a great example of a projection where the shape (or angles) of the countries were vital so the area was distorted - quite dramatically in some places. For example, have a look at Greenland and Africa. In the Mercator projection, the look about the same size - but this is just not the case in reality. The globe shows their relative size much more clearly. The *Peters projection* (figure 2.1, on the facing page) is the opposite example - the area was maintained but the shape was heavily distorted. Have a look at Africa and Greenland again. Most of the time for global maps, some type of compromise is used where shape and area are distorted a bit but not

---

[7]Technically, it is not quite a sphere as it is wider around the equator and shorter from pole to pole, a bit like a squashed sphere. For more details see [9] section 2.4.1 and [11].

too much. One of the most common projections used for global data is the Equal Earth projection. Check out geo-projections.com for more examples of different projections.

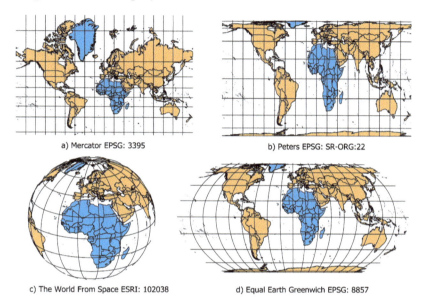

a) Mercator EPSG: 3395

b) Peters EPSG: SR-ORG:22

c) The World From Space ESRI: 102038

d) Equal Earth Greenwich EPSG: 8857

Figure 2.1: The world shown using different projection systems a) Mercator, b) Peters, c) Globe and d) Equal Earth Greenwich.

Note how in figure 2.1 the relative size and shape of Africa & Greenland (in blue/darker grey) vary. The only projection that shows them in their true size is c) the globe view.

Map created in QGIS, data from Natural Earth, projections specified for each map. The globe was created using a updated version of "The World from Space", see statsmapsnpix.com for details.[8]

Projections are really important for global maps, but less important for maps of small areas. Projections are very closely related to *coordinate systems*, which are how we use numbers to represent physical location.

---

[8]http://www.statsmapsnpix.com/2019/09/globe-projections-and-insets-in-qgis.html

2.3.2   Coordinate Systems

Coordinate systems are the fundamental system we use to represent locations using a pair of numbers. One widespread coordinate system, the geographic coordinate system, uses degrees (°) for these numbers. With this, the first number represents a distance north or south from the line 0°, which you probably already know as the equator. This is *latitude*. The second number represents a distance east or west from 0°, the prime meridian (which runs through Greenwich, London, UK). This is *longitude*.

For global maps we have two planes of direction, east/west and north/-south, each with 360°. So, for a global coordinate system, we use angles, with 360° covering the entire globe: +180° to -180° (east/west of the prime meridian all the way round to the international date line in the Pacific Ocean), and 0° to 90° (from the equator to the north/south poles). This is what we call a *geographic coordinate system*; it covers the whole globe, and we can use it to represent any location on the earth's surface using angular units - degrees (°).

There are many different geographic coordinate systems. One of the most common is latitude/longitude, or sometimes called *WGS 1984*. This is the most common one for global data, and has the EPSG code number 4326. An example is 52°N 37' 30.32", 1°E 14' 2.05". The subdivisions are minutes (60 in a degree) and seconds (60 in a minute). You can also represent this as 52.6250°, 1.2339°, which is known as *decimal degrees*.

Decimal degrees is a much easier number for a computer to handle. You will usually get coordinate data in this format rather than separated into degrees, minutes, and seconds. Geographic coordinate systems always use angular units, such as degrees, but this makes measuring stuff tricky. The size of a degree varies depending where on the globe you are. You can work it out, but it requires some complex trigonometry to do this.

EPSG Codes

Every coordinate system is identified by a 4 or 5-digit code. Latitude/longitude is 4326, UTM Zone 30N is 32630, British National Grid is 27700, and there are many others. The EPSG system was set up by the European Petroleum Survey Group, which is where the acronym came from. The organisation no longer exists (the OGP Surveying and Positioning Committee now runs it), but the acronym has stuck. All GIS use these codes to record which system a set of spatial data is using, and R makes heavy use of this code system. When we reproject data (move from one system to another), we have to specify which system we want R to reproject the data to, using the EPSG code for that system.

We have another set of coordinate systems - *Projected Coordinate Systems*. These are always based on specific projections (hence the name) and cover a small area (usually a maximum of 6° longitude / about 600km and from the equator to 80°N). The projection used for each of these assumes the area is flat, and ignores the curvature of the earth. This is what limits the size of each projected coordinate system.

Small countries, such as the United Kingdom, are usually covered by one projected coordinate system. Large countries, such as Brazil, will typically be covered by several. The big advantage is that the units used by these systems are linear (usually meters, but sometimes feet in the US) rather than angular. This means it is very easy to measure distances using coordinates and the formula is very simple using Pythagoras theorem, $a^2 + b^2 = c^2$, if you remember that!

There are two common projected coordinate systems. *Universal Transverse Mercator (UTM)* is global. The UTM system splits the globe up into a series of grids, and each grid is a projected coordinate system. You then have a coordinate (X & Y meters) from a nominal 0, 0 point. Each grid has a different EPSG code. Each grid is usually 6° longitude and from the equator to 80°N. The grids are numbered (zones 1 - 60) and are either N (north of the Equator) or S (south of the equator).

The other projected coordinate system, if you're working in the United Kingdom, is the *British National Grid (BNG)*. This covers the whole of the UK and uses a pair of numbers to represent location, using metres.

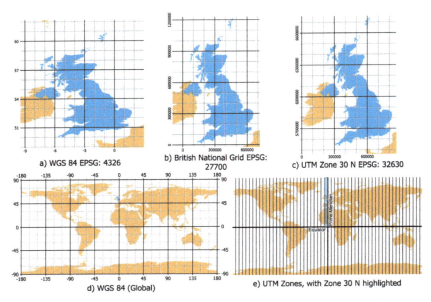

Figure 2.2: Different common coordinate systems that cover the UK: a) WGS 84 (sometimes called latitude/ longitude) EPSG: 4326, b) British National Grid EPSG: 27700, c) UTM Zone 30 N EPSG: 32630; d) Global WGS 84 and e) the UTM Zone system, with Zone 30 N highlighted.

UTM Zones and the Military Grid Reference System

There is a key subtlety in the UTM Zone system that it is worth being aware of. The zone system is the 6-degree bands that run from the North Pole to the Equator (N) and the Equator to the South Pole (S). There is an additional grid system (formally called Military Grid Reference System) that uses the same zones (1-60) but then uses the letters C - X to further split the bands into 8-degree grids. Confusingly, they include the letters N and S in their C - X range. There is a very good example of the confusion this can cause here.[a] In short, if your letter is N or S, it is probably in the Zone system, if you have other letters (e.g. 30 U) then it is probably in the Military Grid Reference System. The zone system is by far the most common. If in doubt, pick one and see if the data appears in the right place!

[a]https://gis.stackexchange.com/questions/162925/

There is much more we could say about coordinate systems, but the key aspect is to make sure you know which one you are using. You can find a selector on the epsg.org website, which will also tell you what code number it is. If you want to find out more, check out Geocomputation with R, the GIS wiki's page on the geographic coordinate system[9], and the ArcGIS page on differences in coordinate systems[10].

2.4 Conceptual Models

One key aspect of working with *spatial data* in any GIS, is that you are working with a model of the real world. Like any models, your model has limitations on what it can be used for. What these limitations are depends on how the data was collected. The limitations also dictate what you can and can't do with the data.

Let us start off with how we conceptualise space. There are two dominant conceptual models that are used in GIS - *object-based* and *field-based*. The object-based approach views data as a series of objects, with space around those objects. These objects are usually represented using *points, lines,* or *polygons*. This way of representing data is termed *vector data*.

Whether you use points, lines, or polygons for a specific object depends on what it is and what *scale* you are looking at (more on scale later in this chapter). A point is a pair of coordinates representing one location. You might use this to represent your current location on a map, a post box, or a point to meet people. Lines are used to represent linear objects such as roads, rivers, or railways. Polygons are used to represent areas - anything from building footprints, to administrative areas, county boundaries, national parks, etc.

Points are the base unit - each point is a pair of coordinates in a specific coordinate system (see the section above for more details on coordinate systems). Lines are then a series of points, joined in a specific order. Polygons are a closed area, created by joining points in a specific order. These points, lines, and polygons can be combined, symbolised, and analysed in different ways to represent GIS data.

---

[9]http://wiki.gis.com/wiki/index.php/Geographic_coordinate_system

[10]https://www.esri.com/arcgis-blog/products/arcgis-pro/mapping/coordinate-systems-difference/

Each *vector layer* can only store one type of data - point, line, or polygon. If you want to use more than one type, you will need more than one layer.

Figure 2.3: The three different types of vector data, a) points, b) lines and c) polygons. Note how these are all discrete objects.

The other conceptual method is the field-based approach. This conceptualises space as a *continuous grid*, covering a set area. This grid has cells. Each cell has the size of the area it represents on the ground, called "resolution". It might have a resolution of 1km, 100m, or even as detailed as 15cm.

Each cell in this grid then has a value which represents some data. This could be height, where the figure represents meters above sea level. It could also be rainfall, sunshine hours, temperature, or population. The numbers might also be categorical, for example, land cover where 1 = urban, 2 = deciduous forest, and 3 = coniferous forest. Satellite imagery also falls in this category, with the numbers in each cell (or pixel) representing the colour either in visual wavelengths (what we can see) or in non-visual wavelengths (such as infrared).

The continuous grid means that every location has a value. This works well for some data types, but not for others. The object-based approach allows for gaps in the data, which is sometimes important. For example, for a road, a vector format would be most appropriate. This is because the line format shows where the road is, and elsewhere there is no road. Using a raster format for a road would not work very well. While it would work, you would have to store empty data for all the cells in the surrounding raster even if there is no road data.

Raster data formats are usually defined with coordinates for the top left corner (again, in a specific coordinate system) and a specific resolution (usually metres, but the unit is based on the coordinate system). Then the GIS can work out where the whole raster layer is located.

a)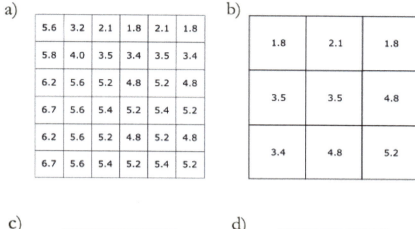

b)

c)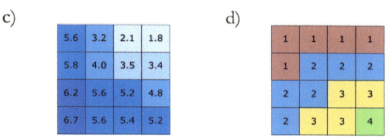

d)

Figure 2.4: Examples of raster data. The resolution is higher in a) and lower in b). Raster data can be used for c) quantitive data and d) qualitative data. Note that this is continuous data - every location (within the raster coverage) has a value.

Also, it is worth mentioning that the object-based data is usually, but not always, vector data. The continuous value data is usually, but not always, raster data. There are some formats that sit between the two, a *vector grid* being one of them. This is a vector format (polygons) but is actually a grid that forms a continuous surface. A *Triangulated Irregular Network* (TIN) also fall within this category.

	Object based	Field based
Vector data	Points, lines, polygons	Vector grid
Raster data	Individual raster cells	Raster grid

Usually you would do all of your GIS analysis with data in entirely one format or the other, either vector or raster. You can convert data from one to the other, but you would usually try to keep this to a minimum because you will lose data quality each time you do the conversion. In social science, vector data is mostly used. In environmental science and physics, raster data is more common.

2.5   Spatial Data File Formats

Alongside the data type (vector or raster) we also have the *file type* - how the GIS data is actually stored on our computer. Some file types can store both vector and raster data, and others can only store one type. Some can store multiple layers in one file, whereas others can only store one layer.

2.5.1   Shapefile

The *shapefile* format is the most common form of GIS data file format. Shapefiles have been around for many years, and they were first introduced in the early 1990s. The fact they've been around for such a long time means that every GIS can read and write shapefiles. They're a really useful way of exporting data from one program to another. Shapefiles are used for vector data, and each shapefile can only store one layer of data. While shapefiles are ubiquitous, they do have some limitations.

Each shapefile is actually made-up of three or more separate files. For example, if you look at a shapefile in Windows Explorer (or Finder), you will see at least three files with the same name but a different extension:

- myshapefile.shp
- myshapefile.shx
- myshapefile.dbf
- myshapefile.prj *(optional)*

The file ending in .shp is the important one - this is the one you open in your GIS. This file contains the actual shapes & coordinates themselves: points, lines, or polygons. The .shx is an index file, and .dbf is the attribute table. The .prj file is optional, and contains the projection and coordinate system information. If you don't have a .prj file, R will not know what coordinate system to use. We will need to specify this manually - more on this later.

When using shapefiles it is really important to make sure you keep all the individual files together. If you don't have at least the three compulsory files, you will not be able to open it. The other key aspects are that sometimes the field names are limited to 10 characters, although this is less of a problem in modern software. Finally, shapefiles don't handle large amounts of data well. It depends a bit on your computer power, but a shapefile of more than 200MB is a bit large, and anything over 1GB probably will not work.

There is a very good Wikipedia article on the shapefile format[11], and the original technical description from Environmental Systems Research Institute (ESRI) is also available online[12].

2.5.2 GeoPackage

Another really key format is *geopackage*. Geopackages are quite new, but solve a lot of the issues with shapefiles. A geopackage is only one file, making it much easier to copy and move. You can also store multiple layers within one geopackage. Vector and raster data can also be stored within a geopackage, making it a very flexible format. There is also no limit on the length of field names.

Geopackages are actually just a database table, packaged up into a file. This means they can work well with large amounts of data. If you do need to store data in a database rather than a file, it is easy to switch from a geopackage to a database.

More details are available on Geopackages at geopackage.org[13], and we will use Geopackages in one of the practical exercises.

2.5.3 Other formats

There are many many other spatial data file formats. *ASCII Grid* and *TIFF* are popular raster ones, but the list of formats is almost endless. R can read-in the vast majority of these formats, so don't worry if you

---

[11]https://en.wikipedia.org/wiki/Shapefile

[12]https://www.esri.com/content/dam/esrisites/sitecore-archive/Files/Pdfs/library/whitepapers/pdfs/shapefile.pdf

[13]https://www.geopackage.org/

come across a format you haven't seen before. You will be able to read it!

See the GIS Glossary for a list of all the GIS terms we've used, including the different file formats.

2.6 Data Types

Along with GIS file data types (vector / raster) and GIS file types (shape-file, geopackage, ASCII Grid, GeoTiff) we also have *attribute data* types. These are the data types contained in the attribute table. Remember, the attribute table for vector data is like a spreadsheet with one column for the geometry: points, lines, and polygons. The rest of the table is made up of columns and rows, with each row representing one feature or vector object: a single point, line, or polygon.

There can be one or more columns, and these columns can contain data. They can contain numbers, such as the population for each object, area of each object, or length. They can also contain names, such as country names, or county names. What these columns can store depends on what data type they use. There are two main data types: string (text) and numeric (numbers). The fact that GIS and R use slightly different terminology to refer to the same thing complicates things!

Data Type	Description	Example
text / string / character	plain text	London
number / numeric	any type of number	23.67
integer	whole numbers	3
real	decimal numbers	45.98
single / double / float	different ways of storing numbers	45.98

The key bit to remember is that if you want to represent the data on a *choropleth map*, the data need to be numbers so that R can work out how to split up the data (see Chapter 4 on Choropleth Maps). If the data is text, then R can't do any calculations on them and can't work out the categories. This should be fairly straight forward, right? It usually is - except when it isn't!

Sometimes data that look like numbers, such as 1,2,3,4,5,6 are actually stored as text. In that case, R will not like them and refuse to draw a map. Fortunately, we can convert them (using as.numeric()) but this is something to be aware of. See the example in the choropleth map section.

2.7 Scale, Level of detail & Generalisation

2.7.1 Scale

Finally, we have the issue of *scale*. This is really important when it comes to working out whether a specific data set is suitable for a specific task. As we covered earlier, spatial data is often captured for a specific use. One key element of this is how much detail is collected. We use scale to specify how much detail is collected.

Scale was originally developed for paper maps, and you may have seen this style designation before: 1:50,000. This means that 1 unit of distance on our paper map represents 50,000 units of distance in the real world. For example, 1cm on a 1:50,000 scale map represents 50,000 cm in the real world - or 500m if you work out the units. Different scale maps will cover different areas.

Scale	Example Area
1:100,000,000 (1:100m)	Worldwide
1:30,000,000 (1:30m)	continent e.g. South America
1:8,000,000 (1:8m)	country e.g. the UK
1:100,000	local area, city e.g. Hull
1:50,000	local area, smaller urban area e.g. Norwich

The scale principle broadly translates to digital data as well - although there are sometimes slight differences due to monitor resolution and size. The key point is that different data will have different amounts of detail. For example, if we're working with some global country boundaries, we might have a nice world map as shown in a) from figure 2.5, on the following page. If we then zoom in to different levels b) - d), we will gradually see less and less detail, until when we're at the local area, where we can't use the data at all.

At the world map scale, this data is around 1:100,000,000. If we zoom in, you can see we gradually lose data. With this data, even at the country level you can see the limited detail. This means we can't use this data for a map of the United Kingdom - the scale is wrong, and we don't have enough detail. For the map of the UK, we would need to find a different data set at a more appropriate scale. Most data will say

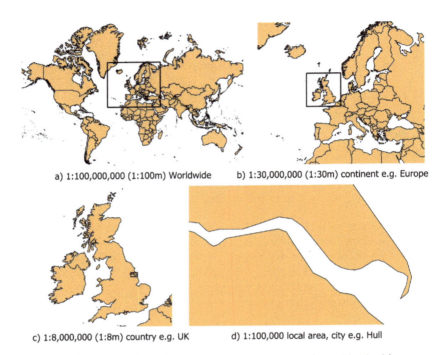

a) 1:100,000,000 (1:100m) Worldwide   b) 1:30,000,000 (1:30m) continent e.g. Europe

c) 1:8,000,000 (1:8m) country e.g. UK   d) 1:100,000 local area, city e.g. Hull

Figure 2.5: Example of one data set at different scales. a) Worldwide, b) continent, c) country, d) local area. Note how the detail is very low in d) so that the map is not usable, and the detail in c) isn't that good - it is debatable whether there is enough detail to use this data at this scale.

(in their metadata) which scale(s) they are suitable to be used at. For example, figure 2.6, on the next page shows that the Ordnance Survey says that the Code-Point Open data can be used at scales from 1:1,250 to 1:10,000.

Ordnance Survey

Code-Point Open

Home > Data Products > Code-Point Open

Code-Point Open
Technical information

Specifications

Use cases	Link statistical and numeric datasets to your postcode data using ONS administrative unit codes and NHS health authority code attributes in each postcode
Data structure	Vector - Points
Coverage	Great Britain
Scale	1:1250 to 1:10,000

Figure 2.6: Technical information for Code-Point Open dataset from Ordnance Survey. Note how it says the data is suitable for use between scales of 1:1,250 and 1:10,000.

> **Large scale and Small scale**
>
> We do have the terms 'large scale' and 'small scale' to refer to different scales, although I'm not a fan of the terms because I can never remember which is which. The relative terms are derived from viewing the scale as a fraction, for example, 1:50,000 being 1/50,000 which is larger than 1:250,000 (1/250,000). However, given that maps are usually the same physical size, a 1:50,000 map will cover a much smaller area than a 1:250,000 scale map, which will cover a much larger area. Often, it is just easier to name the specific scale you're talking about than talking about 'large scale' or 'small scale' maps.

2.7.2   Dynamic Data

So far, we've been talking about data that is at a fixed scale - which is very typical of most of the data we will be using in a GIS. This is data we've downloaded and then opened and evaluated within our GIS. These are what we might call *static data*. When you zoom in on them, you don't get any extra detail.

We do have another sort of data that we can use in GIS, which is commonly used for *basemaps* - something we can call *dynamic data*. When we load in this data, it pulls data in from the internet and shows a version appropriate to the scale we're looking at. When we zoom in on a specific area, it will download more data and show us more detail.

In QGIS, we can load basemaps using the XYZ Tile plugin, and in R we can load a basemap by using the View mode in tmap (rather than the plot mode). Behind the scenes, we're actually using an API to access the data - more on this in Chapter 6.

2.7.3   Level of Detail & Generalisation

Another concept, closely related to scale, is "level of detail" and "generalisation". So far, we have fixed data at a set scale - meaning that if we zoom in, we don't get additional data. Data at a specific scale has a specific amount of information in it, a specific level of detail. The amount of detail in a particular data set is important because it very closely relates to the size of the dataset and the amount of space it takes up on your computer.

The larger the file size that the data set has, the more resources our computer needs to process it. To give you some general ideas of typical sizes for large and small files, a spatial data file size of about 2-3 MB is fairly small, about 15-20 MB is medium, and 60 MB or more is getting large, with anything over 100 MB very large. Of course, what your computer can manage will vary a bit with your computer specification. However, if things are starting to run more slowly than normal, this is an indication that the file sizes might be too big.

R can handle spatial data larger than 100 MB, but things might start to slow down a bit. This is exacerbated if you're dealing with multiple layers. For example, if you have a layer that is 150 MB in size, and you're dealing with 6 or 7 different layers each that size, then they start to add up very quickly. If a specific spatial data set at a specific scale is stored in a shapefile, then the file will be larger than if the same data is stored in a geopackage. Therefore, if you're working with larger data storing it in a geopackage can help.

A full level of detail is often not required for much of the GIS work that you do with a specific data set. We have what is called *generalisation*, where additional detail that is not needed can be removed. Here are some examples for small areas (LSOAs) in the Weymouth area at different generalisation levels.

Level of generalisation	England	Weymouth
Full resolution	457 MB	2.8 MB
Generalised	36.5 MB	231 KB
Super generalised	12.3 MB	99 KB

Typically, there is a balance between size and detail, and for most analysis, the generalised version of the data will be fine. Figure 2.7, on page 47 shows a comparison of the detail that is removed at the different generalisation levels - and what it looks like at different scales. Another term for the same concept is simplification. This is applicable to both line and polygon data sets.

This is a really useful technique to reduce the file size of spatial data, which also makes it much quicker to perform spatial analysis on the data.

We do have to be careful to balance the amount of simplification we do - there still needs to be enough detail in the data for whatever analysis we want to do later on, and of course what level this is depends on a) the data and b) the analysis we want to do later on.

For example, in the bottom row of figure 2.7, on the next page, we've reduced the detail too much. This might have impact on the analysis we do - for example, if we did a point in polygon analysis on the super-generalised data, some points might end up in the wrong polygon.

2.8   What Can We Use GIS For?

GIS has many different uses across the board. It can be a really useful tool in any area where location is important - and I would say that is pretty much anything! There is a great illustration, The Spatial Analysis Periodic Table, put together by GIS Geography that lists all the different tools in GIS and the areas it could be used.[14] There are many! They split the application list into Nature/Environment and Society/Economy.

---

[14]Check it out at https://gisgeography.com/spatial-analysis-periodic-table/

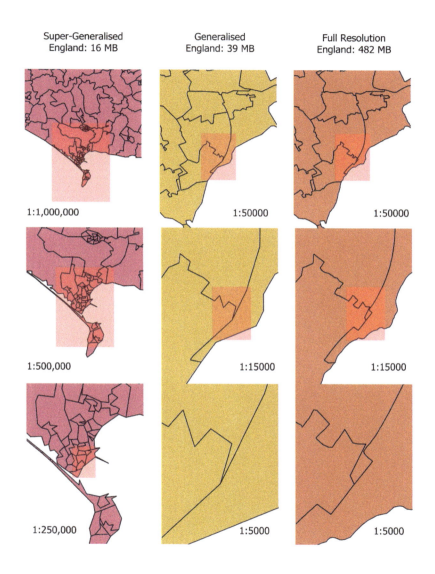

Figure 2.7: An example of data simplification using Census based LSOA (Lower layer Super Output Area) boundaries for Weymouth, on the south coast of England. Each column has a different level of generalisation, going from most generalised (on the left) to least generalised (on the right). The data is shown at different scales so you can see when the data start to lose detail. Also notice the different file sizes for each level of generalisation.

3. Using R as a GIS

3.1 Learning Outcomes

After reading this chapter, you will:

- Understand how to use libraries in R.
- Know how to load spatial data in R.
- Understand how to join data.
- Understand how to draw a quick map using the qtm() function.

Having covered the background to GIS (in Chapter 2) and R (in Chapter 1), now we can combine them. In this chapter we will look at reading spatial data into R and understanding what it shows and how it is structured (using the sf library). We will also read in a CSV file and join the two together. Finally, we will create some basic maps using it. We're going to focus initially on vector data, with raster data being covered in Chapter 5.

3.2 Spatial Data Libraries

First of all, we need some libraries to work with spatial data. Libraries allow us to add extra features to R, such as GIS tools and analysis. There are two stages to working with libraries - installing the libraries, and loading the libraries. We need to install the library only once on any computer that we use R, but we need to load the library every time we start RStudio.

The installation goes through the process of downloading the files from the internet and making sure they're all there for when R wants to use them. Think about installing a light bulb - to install it, you need a light fitting, and you need to attach the bulb to the fitting. The R code for this is install.packages("package name").

Packages and Libraries

Helpfully R uses both the term package and the term library. They mean the same thing and can be used interchangeably when talking about them. However, the installation function is install.packages() (there is no install.functions()) and the load function is library() (there is no function package()!) Unfortunately this type of inconsistency in R is not unusual.

We have two libraries we need to install: sf and tmap.

For sf run this code:

```
install.packages("sf")
```

Once it is installed, we need to load it. This tells R we're actually going to use it today. Going back to our light bulb analogy, we need to press the switch to turn the light on. The code for this is:

```
library(sf)
```

When you run this, R will think about it for a short time, and if everything has worked, it will go to the next line on the console, >.

We also need to do the same process for another library - tmap. See if you can work out the code to do this. If you get stuck, the code is in the ch3-script.R file. Remember, tmap is quite a big library, so be patient - it can take up to 5 min for it to fully install, and there may be a min or two where R looks as though it isn't doing anything.

> **tmap v4**
>
> When you install tmap in R, you get version 3.3-4. When you load the library, you will probably get this message:
>
> ```
> > library(tmap)
> Breaking News: tmap 3.x is retiring. Please text v4, e.g.;
>         with remotes::install_github('r-tmap/tmap')
> ```
>
> This is saying that there is a new version of the tmap library - version 4. It is not out yet and RStudio will have installed version 3.3-4 for you. We will stick with using version 3 in this book, and check out section 4.10 (tmap v4) on page 87 for more details.

3.2.1 Troubleshooting

There will be times R will not be happy installing a library for some reason. Sometimes the error message it gives will be helpful. It might say something like this:

```
There has been an error installing the Rcpp library
```

It will probably reference a specific package, like the error above does. A good approach is to try just installing this package on its own. Often packages have dependencies, which means it relies on other packages, so when you install sf, R also automatically installs classInt, graphics, grid, Rcpp and a bunch of others. Sometimes these others also have dependencies, so you can end up with quite a long list!

With the example above, R has had a problem installing Rcpp, so try installing just that:

```
install.packages("Rcpp")
```

Then try installing sf and tmap again.

The other thing to try is IT 101: restart RStudio and try again. Quite often this does solve the problem, so it is worth trying. If that doesn't work, you can also try restarting your computer.

> If you have problems installing the libraries, check the error messages. Sometimes it will mention a dependent library. If you install the dependent library and then install the main library, it will work. It is also worth trying either restarting RStudio, or your computer, to see if that solves the problem. You can also find answers on my online guide for installing RStudio[a] under *Troubleshooting* and in Chapter 16.
>
> ───────────
> [a]`https://nickbearman.github.io/installing-software/r-rstudio`

3.3 Working Directory

The other key aspect to remember is that R uses working directories. This is where R looks for files when you load them, and if R can't find your file there, it won't be able to load it. Typically, you would have a separate directory for each project you use.

Create a new folder / directory for your work from this book. It doesn't matter where you create it, as long as you know where it is.

There are a couple of ways of setting your working directory. You can use the graphic interface in RStudio shown in figure 3.1.

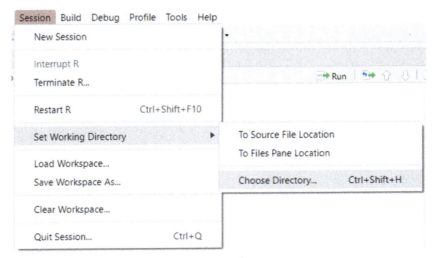

Figure 3.1: Setting a working directory using RStudio.

Or you can set your working directory using code:

```
setwd("~/work")
```

Typically I use the graphic interface as I find it easier!

3.4   Reading in a CSV file

Once we've set our working directory, we can load in a file from it. Download the ch3-ch4-data.zip file from the book website, and save it into your working directory.

This is a zip file, which is a series of files combined together, to make it easier to download. This is very common with shapefiles, as they are made up of 3 (or more) separate files that need to be kept together. You need to extract the files - do this on Windows by double-clicking and choose **Extract All** (you may not need to click Extract All if you are on a Mac).

This will extract the files into a folder, but make sure you check if they've extracted in to a subfolder - if they have, you need to move them to your working directory. If they're in a subfolder of your working directory, R will not be able to find them.

Once this is done, run this command to read in the CSV file:

```
hp.data <- read.csv("hpdata.csv")
```

This command uses the function read.csv(). It reads in the file "hpdata.csv" from the working directory and saves it as a variable called hp.data.

> The . (period) in hp.data is just part of the name. We could have equally called it hpdata, hp_data or something else. What would not work is including spaces: hp data would not be a valid variable name.

We can use the head() function to check that the data was read in properly.

```
head(hp.data)
```

This should output:

```
ID Burglary Price
1 21        0   200
2 24        7   130
3 31        0   200
4 32        0   200
5 78        6   180
6 80       19   140
```

Now that we have some non-spatial data in R, let's bring in some spatial data as well.

3.5   Reading in and Mapping a Shapefile

Along with the CSV file, we also had a shapefile in our zip file. Run this command to read it in:

```
library(sf)
sthelens <- st_read("sthelens.shp")
```

This command uses the function st_read(). It reads in the file "sthelens.sh (from the working directory) and saves it as a variable called sthelens.

> The file name and the variable name aren't related. We could have used this code: banana <- st_read("sthelens.shp") which would call our variable banana. But it is important to have clear variable names, and often makes sense for them to be the same as the file name - which should explain what information is stored in that variable.

It should give you the output below:

```
Reading layer 'sthelens' from data source

'C:\Users\nick\Documents\GIS\using-r-gis\sthelens.shp'
  using driver 'ESRI Shapefile'
Simple feature collection with 118 features and 5 fields
Geometry type: POLYGON
Dimension:     XY
Bounding box:  xmin: 345350.2 ymin: 387836.6 xmax: 361798.8 ymax: 404145.6
Projected CRS: Transverse_Mercator
```

You can also see that sthelens is listed in your Environment tab (top right of the RStudio window) (see figure 3.2).

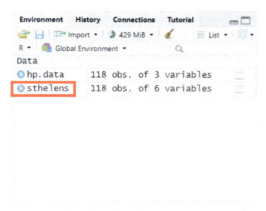

Figure 3.2: sthelens now listed in the RStudio environment (in the top-right of your RStudio window).

St. Helens is now stored as a "Simple Features" (or sf) object. You can use the qtm() function to draw the polygons (i.e. the map of the LSOA).

```
library(tmap)
qtm(sthelens)
```

The qtm() function is Quick Thematic Map, and is a quick way of getting a map output. It shows us that we've read the data in ok, and what the data looks like (see figure 3.3, on the following page).

Along with the spatial element of vector data, each file has an attribute table. This is a table of data, with each row linked to a specific polygon. We can use the head() command to view the first six rows of the attribute table.

```
head(sthelens)
```

See output in 3.4, on the next page.

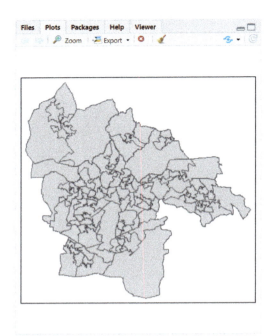

Figure 3.3: Output in the Plots window from qtm(sthelens)

```
Simple feature collection with 6 features and 5 fields
Geometry type: POLYGON
Dimension:     XY
Bounding box:  xmin: 345350.2 ymin: 387836.6 xmax: 356013.9 ymax: 403054
Projected CRS: Transverse_Mercator
ID  GID        NAME         LABEL  ZONECODE    geometry
1 21 2049 St. Helens 009D 04BZE01006825 E01006825  POLYGON ((348074.3 395549.9...
2 24 2052 St. Helens 001B 04BZE01006883 E01006883  POLYGON ((347824.8 401169.7...
3 31 3564 St. Helens 022A 04BZE01006898 E01006898  POLYGON ((351561.1 390368, ...
4 32 3565 St. Helens 001D 04BZE01006885 E01006885  POLYGON ((351662.3 402905.2...
5 78 4389 St. Helens 023C 04BZE01006891 E01006891  POLYGON ((348260 391439, 34...
6 80 4391 St. Helens 018C 04BZE01006828 E01006828  POLYGON ((348188.1 392976.1...
```

Figure 3.4

This is the same as the attribute table in programs like ArcGIS, QGIS or MapInfo. If you want to open the shapefile in QGIS or ArcGIS to have a look, feel free to. Shapefile is a standard GIS format, and can be opened by any GIS program.

As well as the attribute table, we can see quite a range of other information. The header information says what type of vector file it is `Geometry`
`type:` `POLYGON` and what projection and coordinate system the data
are in `Projected CRS: Transverse_Mercator`.

We also then have the columns themselves - `geometry` is the coordinates
of the polygons. There are a few others, including `ID` which is the one
we're interested in.

We're going to join this data to the `hp.data` variable. Both data sets have
the ID column in them, and this is what we're going to use to join them
(Figure 3.5).

```
> head(sthelens)
Simple feature collection with 6 features and 5 fields
Geometry type: POLYGON
Dimension:     XY
Bounding box:  xmin: 345350.2 ymin: 387836.6 xmax: 356013.9 ymax: 403054
Projected CRS: Transverse_Mercator
  ID  GID            NAME        LABEL  ZONECODE                    geometry
1 21 2049 St. Helens 009D 04BZE01006825 E01006825 POLYGON ((348074.3 395549.9...
2 24 2052 St. Helens 001B 04BZE01006883 E01006883 POLYGON ((347824.8 401169.7...
3 31 3564 St. Helens 022A 04BZE01006898 E01006898 POLYGON ((351561.1 390368, ...
4 32 3565 St. Helens 001D 04BZE01006885 E01006885 POLYGON ((351662.3 402905.2...
5 78 4389 St. Helens 023C 04BZE01006891 E01006891 POLYGON ((348260 391439, 34...
6 80 4391 St. Helens 018C 04BZE01006828 E01006828 POLYGON ((348188.1 392976.1...

> head(hp.data)
  ID Burglary Price-thousands
1 21        0             200
2 24        7             130
3 31        0             200
4 32        0             200
5 78        6             180
6 80       19             140
```

Figure 3.5: The head() of `sthelens` and `hp.data`, showing the `ID` field which we
are going to use for the join

This is what we call an attribute join. This is where we use a common
attribute (between the non-spatial and spatial data sets) to link the two
data sets together. In our example here, we have an ID number stored
in the `ID` field. With other data, it might be a common name (e.g. the
name of a country) or some other type of ID number.

The key element is that the attribute has to be identical. If we were using country names, if we had in the spatial data set "United States" and
in the non-spatial data we had "United States of America", the computer would not be able to join these successfully. Even though we can
clearly tell they're referring to the same thing, the computer doesn't

realise this, so they have to be identical as a computer would view it.

The merge function joins the data. In this case, R is clever enough to look at both data sets, see there is a field called ID in both datasets and join them automatically. If the fields weren't called the same thing, we would need to specify the fields - and we will do this in a later example.

```
sthelens <- merge(sthelens, hp.data)
```

We can use the head function to check that the data have been joined correctly:

```
head(sthelens)
```

See output in 3.6.

```
Simple feature collection with 6 features and 7 fields
Geometry type: POLYGON
Dimension: XY
Bounding box: xmin: 346795.9 ymin: 389514.8 xmax: 351909.3 ymax: 404145.6
Projected CRS: Transverse_Mercator
ID GID NAME LABEL     ZONECODE                     Burglary Price... geometry
1 100 4411 St. Helens 001C 04BZE01006884 E01006884     0      180   POLYGON ((348058.5 40072...
2 101 4412 St. Helens 001F 04BZE01006887 E01006887     0      200   POLYGON ((347413.4 40132...
3 102 4413 St. Helens 001E 04BZE01006886 E01006886    15      210   POLYGON ((348792.3 40316...
4 111 6849 St. Helens 023B 04BZE01006889 E01006889     6      170   POLYGON ((350108.1 39019...
5 112 6850 St. Helens 022E 04BZE01006910 E01006910    12      180   POLYGON ((351675.1 39142...
6 113 6851 St. Helens 019H 04BZE01006913 E01006913     8      160   POLYGON ((350807.5 39328...
```

Figure 3.6

Now we can see how the data has been joined.

Finally for this chapter, we can use the qtm() function again to draw a quick map, to show the burglary data (Figure 3.5):

```
qtm(sthelens, fill="Burglary")
```

This shows the data for us, and is a quick way to generate a map. We don't have any way of customising it, but we can use a different function for this - tm_shape() - which we will look at in the next chapter.

You can export the map if you want to: click on the *Export* button, and then choose *Copy to Clipboard....* Then choose *Copy Plot.* If you also have Microsoft Word up and running, you can then paste the map into your document. You can also save the map as an Image or PDF. There

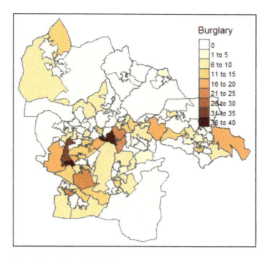

Figure 3.7: Burglary data for St Helens

are also other ways of exporting maps from R, which we will come to in due course.

Part 2

Spatial Data

4. Choropleth Maps Using Vector Data

4.1 Learning Outcomes

After reading this chapter, you will:

- Understand what a choropleth map is.
- Know the decisions we need to make when classifying data and creating a choropleth map.
- Understand the different classification methods and how they impact the resulting map.
- Be aware of the issues to consider when choosing colour schemes.
- Understand how the type and characteristics of different data sets impact these decisions.

This chapter introduces a new type of map called choropleth maps, which is a common method used to display aggregate data on a map. We will discuss colour and classification choices for choropleth maps, including how we can create and customise them using the tmap library. We will also talk about types of data (quantitative, categorical, qualitative) and how these are best represented.

4.2 Creating a Choropleth Map

Choropleth maps are one of the most common form of maps, where we use colour to show data. This can be at a range of different geographic levels - e.g. counties, states or even countries. A typical example is shown in figure 4.1, on the following page.

We have a range of different options to choose when we create a choropleth map, which we will cover in this chapter. First of all, though, we need to get the data into R to create a map.

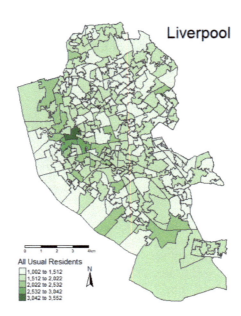

Figure 4.1: A typical choropleth map showing the number of people resident in different output areas in Liverpool.

There are two key bits of data - the spatial data, showing whatever geographies we want to use. Here we're going to use a shape file of the countries of the world as an example. First of all, we need to read it in:

```
library(sf)
countries <- st_read("world_countries.shp")
```

As before, R will give you some information:

```
Reading layer `world_countries' from data source
  `C:\Users\nick\Documents\GIS\using-r-gis\world_countries.shp'
  using driver `ESRI Shapefile'
Simple feature collection with 230 features and 11 fields
Geometry type: MULTIPOLYGON
Dimension:     XY
Bounding box:  xmin: -180 ymin: -90 xmax: 180 ymax: 83.6236
Geodetic CRS:  WGS 84
```

We can also plot the map as shown in figure 4.2.

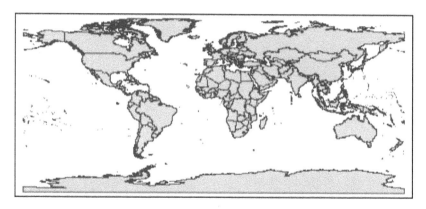

Figure 4.2: A typical choropleth map showing the number of people resident in different output areas in Liverpool.

Secondly, we need the data we're going to map. If we look at our world country data, using the head() function, we can see it has some data in it:

```
head(countries)
```

```
Simple feature collection with 6 features and 11 fields
Geometry type: MULTIPOLYGON
Dimension:    XY
Bounding box: xmin: -61.89111 ymin: -18.01639 xmax: 50.37499 ymax: 42.66194
Geodetic CRS: WGS 84
  FIPS ISO2 ISO3 UN              NAME  AREA POP2005 LIFEEXP LIFEEXPM
1   AC   AG  ATG 28 Antigua and Barbuda    44      83   75.00    72.56
2   AG   DZ  DZA 12            Algeria 238174   33268   73.06    71.02
3   AJ   AZ  AZE 31         Azerbaijan   8260    8563   70.12    67.09
4   AL   AL  ALB  8            Albania   2740    3082   76.30    73.42
5   AM   AM  ARM 51            Armenia   2820    3015   74.00    70.58
6   AO   AO  AGO 24             Angola 124670   17913   49.63    48.21
  LIFEEXPF  INFMRT              geometry
1    77.36  10.028 MULTIPOLYGON (((-61.73806 1...
2    75.22  34.176 MULTIPOLYGON (((2.96361 36....
3    73.14  41.062 MULTIPOLYGON (((45.31998 39...
4    79.72  16.108 MULTIPOLYGON (((19.43621 41...
5    77.30  21.009 MULTIPOLYGON (((45.21137 41...
6    51.04 104.349 MULTIPOLYGON (((11.775 -16....
```

We have a range of columns - `NAME`, `AREA`, `POP2005`, `LIFEEXP`, `LIFEEXPM`, `LIFEEXPF` and `INFMRT`. Initially, we're going to show the population data. However, the 2005 data is a little bit old, so we're going to join on some more up-to-date data, from 2015. This data is in `pop2015.csv`, which you downloaded earlier.

Read the data into R

```
pop2015 <- read.csv("pop2015.csv")
head(pop2015)
```

Commas (,) and full-stops (.) in Comma Separated Value (CSV) files

English, and a number of other languages, use a '.' as a decimal point, and sometimes use ',' as a thousands separator, for example: 1,234,567.89. However, a number of European languages (and others) use ',' as a decimal point and occasionally '.' as a thousands separator, for example 1.234.567,89. When working with data typically thousand separators are not used, so we can ignore those. However decimal points are important for numbers, and also important for CSV files. CSV stands for Comma Separate Values, and a typical CSV file is structured like this: (open world-cities.csv in a text editor if you fancy a look):

```
Name,Latitude,Longitude,Country
Shanghai,31.23,121.47,China
Bombay,18.96,72.8,India
Karachi,24.86,67.01,Pakistan
```

You might be able to see where we are going with this! If we were using a comma as a decimal point, we wouldn't be able to tell apart the separate columns from the numbers with a decimal point ','. Instead, what usually happens is that ; are used instead of , as a separator in this instance:

```
Name;Latitude;Longitude;Country
Shanghai;31,23;121,47;China
Bombay;18,96;72,8;India
Karachi;24,86;67,01;Pakistan
```

We would normally use read.csv() function to read CSV data in. However it is expecting a comma ',' separator and if it gets a semi-colon ';' the function won't work. So read.csv2() was created which does read in ';' semi-comma. Otherwise the functions are identical. (See stackoverflow.com/questions/22970091 for a good explanation).

What complicates matters is if you use Excel to save a CSV file, the choice of separator is influenced by the local language setting of the computer you are using. If you're not sure, try both and see what happens. If you want to check, you can open the CSV file in a text editor (e.g. Notepad) NOT Excel, and you can see how the file is structured.

Once we have read in the file, we can use the head() function to see the first 6 rows:

```
head(pop2015)
```

```
  UN_Code               Name POP2015
1      28 Antigua and Barbuda      92
2      12                Algeria   39667
3      31             Azerbaijan    9754
4       8                Albania    2897
5      51                Armenia    3018
6      24                 Angola   25022
```

Now we have the data we want to map, but we need to join it to the spatial data, so that R knows where each country is. Again we can use the merge() function, like we did with the burglary data earlier on. However, as the fields we're joining this time have different names, so we need to specify them explicitly:

```
#join pop2015 data to countries data
countries <- merge(countries, pop2015, by.x="NAME", by.y="Name")
```

Here the order of the variables is key. Note that the countries variable containing the spatial data is first, and pop2015, the attribute data, is second. The by.x is the column name for the first data set (countries) and the by.y is the column name for the second data set (pop2015).

Run this code, and then to check that R has done what we want to, do use head(countries):

```
head(countries)
```

```
Simple feature collection with 6 features and 13 fields
Geometry type: MULTIPOLYGON
Dimension:     XY
Bounding box:  xmin: -170.8261 ymin: -18.01639 xmax: 74.91574 ymax: 42.661
Geodetic CRS:  WGS 84
            NAME FIPS ISO2 ISO3 UN   AREA POP2005 LIFEEXP LIFE
1    Afghanistan   AF   AF  AFG  4  65209   24400   58.04   56.96
2        Albania   AL   AL  ALB  8   2740    3082   76.30   73.42
3        Algeria   AG   DZ  DZA 12 238174   33268   73.06   71.02
4 American Samoa   AQ   AS  ASM 16     20      59      NA      NA
5        Andorra   AN   AD  AND 20      0      81      NA      NA
6         Angola   AO   AO  AGO 24 124670   17913   49.63   48.21
  LIFEEXPF  INFMRT UN_Code POP2015                       geometry
1    59.20  79.509       4   32527 MULTIPOLYGON (((74.91574 37...
```

```
2    79.72  16.108      8   2897 MULTIPOLYGON (((19.43621 41...
3    75.22  34.176     12  39667 MULTIPOLYGON (((2.96361 36....
4      NA      NA       16     56 MULTIPOLYGON (((-170.7494 -...
5      NA      NA       20     70 MULTIPOLYGON (((1.78172 42....
6    51.04 104.349     24  25022 MULTIPOLYGON (((11.775 -16...
```

Have the two data frames joined correctly? While the `head()` function only shows us the first six rows, we can look at the whole data set at once in RStudio, using the `View()` function. (Note the capital V, R is case-sensitive!). As shown in figure 4.3

```
View(countries)
```

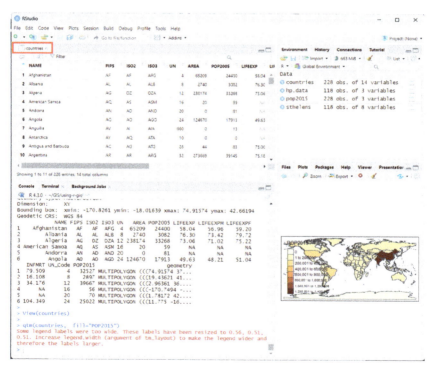

Figure 4.3: RStudio, showing the countries variable using the `View()` function.

If you're working with a large dataset, this might take a while to open - so be patient.

This gives us an interactive table we can scroll around. You can also reorder the rows by clicking the up/down arrows at the top of each column, to sort in ascending or descending order.

We can also use the qtm() function that we used in earlier chapters. Try it out, to see if you can get the output shown in figure 4.4.

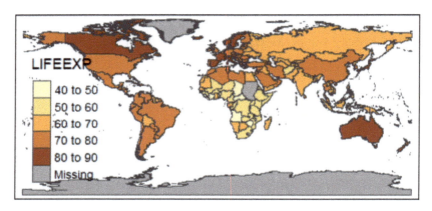

Figure 4.4: Output from qtm() showing the life expectancy data.

The answer is in the *ch4-choropleth-maps-script.R* file on the book website. Try not to look at the answer until you have had a go yourself!

4.3   Using Scripts in RStudio

So far you've probably been typing your code directly into the console - the section at the bottom left of the screen in RStudio. This is fine, but as the code gets more complex, you might need to go back and edit it. You've also probably been copying from the book, which is also fine, but if you want to go back and check content, it gets tricky.

This is where scripts come in. We can save our code in a file, and RStudio will allow us to edit this, and run bits of code as we wish to. To start one, choose:

- **File > New File > R Script** (see figure 4.5, on the facing page)

A new, blank script will open with the name *Untitled1*. You can type your commands in here and then run them from here. Type in one of

Figure 4.5: Click **File** > **New File** > **R Script** to create a new script.

the commands we used earlier:

```
pop2015 <- read.csv("pop2015.csv")
```

Just typing the code in doesn't make RStudio run it. To do this, click anywhere on this line, and click **Run**, or use the keyboard shortcut, hold down Control (sometimes labelled Ctrl) and press Enter (written as Ctrl + Enter). See figure 4.6.

Figure 4.6: Running code from a script, using the **Run** command or keyboard shortcut **Ctrl + Enter**.

You will see the code appear in the console below.

Add in the rest of the code we have already done, and it should look
something like this:

```
## Chapter 4: Choropleth Maps

setwd("~/GIS/using-r-gis")

# Read in a csv file

pop2015 <- read.csv("pop2015.csv")
head(pop2015)

# Read in a shapefile

library(sf)
countries <- st_read("world_countries.shp")

library(tmap)
qtm(countries)

head(countries)

#join pop2015 data to countries data
countries <- merge(countries, pop2015, by.x="NAME", by.y="Name")

head(countries)

View(countries)

qtm(countries,  fill="POP2015")
```

It is not compulsory to add all this in, but it is good practice. The idea
is if you open a script, you should be able to run it from the beginning
and it will do all the processing you need.

Now, when we add more code, add this to your script.

4.4 Colours

We've already some some bits with the qtm() function, but if we apply
this to our newly joined Pop2015 data, we don't get a great map (4.7, on
the facing page):

```
qtm(countries,  fill="POP2015")
```

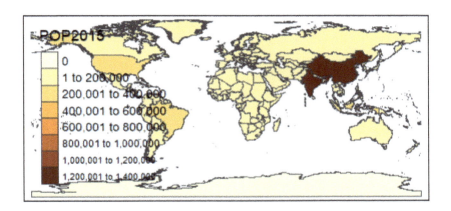

Figure 4.7: Running code from a script, using the **Run** command or keyboard shortcut **Ctrl + Enter**.

There are a few things wrong with it:

- The colours don't really work - we have two countries (China and India) in the top category in dark brown, three in the middle brown category (USA, Brazil and Indonesia) and the rest all in one colour.
- There is a category with a value of 0 that doesn't mean much.
- The legend is covering sections of the map.

I'm sure there are many more problems too!

The qtm() function is quite limited, but we can address some of the issues by using the tm_shape() function instead.

Initially, we can also use the tm_shape() function, to get a basic map:

```
tm_shape(countries) +
  tm_polygons("POP2015")
```

Currently, this is the same map as we had with qtm() function, but we have a lot more options with the tm_shape() function. We can customise the title, colours, classification and a whole range of other options:

```
tm_shape(countries) +
  tm_polygons("POP2015", palette = "Greens", style = "jenks")
```

This gives us some better groups for the data, and now we can see a bit of variation on the map.

We have added two new parameters here - palette = "Greens" and style = "jenks". palette = "Greens" sets the colours.

Try running the same code, but substitute Blues for Greens.

The parameter style = "jenks" sets what classification method we use to split the data up - more on this in the next section!

4.5 Classifications

Swapping out Greens for Blues in the map above makes the map look quite different. When we make a choropleth map, the choices we make in the design are very important, as they have a significant impact on the map. We have three different aspects to think about - the colours we use, the number of classes (or groups) we split the data into and where those classes are actually positioned.

The locations of the classes are probably the most important here, as these can have the most impact on the map. As a quick example, think about this table of data:

Country Name	Forest Percent Cover 2015
Burundi	5.9
Kenya	6.2
Uganda	18.4
Rwanda	19.5
Tanzania	39.9
Angola	47.4
Zambia	57.1
Congo DR	58.9
Congo	65.8
Gabon	84.5

If I gave you a copy of the map (figure 4.8, on the next page), and some coloured pencils, how would you colour it in?

How would you split these up to categorise the data? Spend a minute or two thinking about it before moving on to the next paragraph.

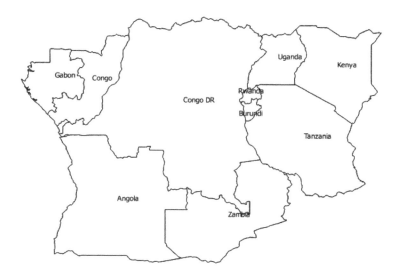

Figure 4.8: Map of 10 countries in Central Africa.

If you have some coloured pencils, try colouring it in! An A4 version called *classification-exercise.pdf* is available in *ch3-ch4-data.zip* on the book website if you want to print it out and have a go yourself.

There are lots of different ways of splitting up this data.

We could take the range of data (0 - 100) and split the data into 5 equal categories, which would be 0-20, 20-40, 40-60, 60-80, and 80-100[15].

We could split it into 10 categories, and have categories 0-10, 10-20, etc., or we could allocate each country to a different category.

We might do something slightly different - we could have 0-10, 10-30, 30-40, 40-50, 50-60, 60+ - as the data aren't equally spread.

There are probably also many other different options we could try - there is no limit really. Whichever approach we use, we will generate a different map - only slightly different in some cases, or wildly different in other cases.

---

[15]Some people would consider this wrong, and say it should really be 0-19, 20-39,

These are the different ways:

Classification Name	Code	Details or Example
Equal Interval	equal	Regular intervals e.g. 0-5, 5-10, 10-15, 15-20
Quantiles	quantile	Split the data into 5 equal categories, with the same number of data points in each category
Natural Breaks	jenks	Algorithm based to create data driven categories
Standard Deviation	sd	Bases classes on data's standard deviation e.g. -2SD to -1SD, -1SD to 0, 0 to 1SD, 1SD to 2SD
Fixed Breaks	fixed	You choose the breaks - see below

This is what the potential maps might look like in figure 4.9.

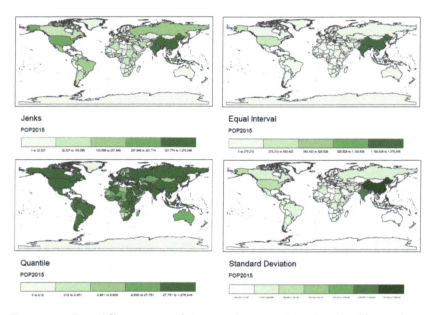

Figure 4.9: Four different ways of showing the same data a) Jenks, b) Equal Interval, c) Quantile and d) Standard Deviation. Note how some make the data much easier to understand, and some make the data much harder to understand.

Each of these will generate a different map. When we create a choropleth map in R using the tmap library, if we used the qtm() function,

40-59, etc. It depends slightly on the data, although technically the groups should not overlap. Formally they would be 0 to <20, (i.e. 0 to less than 20), 20 to <40, 40 to <60 etc. Different groups have different styles so it might be that your map needs to follow a certain in-house style.

it will automatically pick a default. This may (or may not) be the best choice for our data. It depends on what the data is, how it is distributed, as well as what we're trying to show. There are lots of choices, but a good rule of thumb is to start with Natural Breaks - this usually gives a good map which represents the data fairly.

The other, closely related choice is how many categories we split the data up into. For a choropleth map, we would normally go for between 5 and 7 categories. This is because it is difficult to distinguish colours for more categories (see the section below on colours. It is also tricky to remember more than 7 categories if we are interested in the actual values of the data (as opposed to the relative values). If we are only interested in relative values (e.g. is this value higher or lower than this value?), then we can use a greater range of colours. However, if we want to identify which of the 5 categories a specific value, then we need to be able to identify which (of the 5) colours it is. The numbers 5 to 7 crop up in many different areas of cognitive psychology and human behaviour [12], see also[16].

The final aspect of choropleth maps we need to consider is which colours we use. If the data we are showing is numeric or ordinal (i.e. they have some order to them) then we would tend to use a colour scale, e.g. a series of different shades of the same colour. Typically, we would use darker shades to represent higher values, and lighter shades to represent lower values.

A great website for choosing colours is ColorBrewer.org. This was developed by Cynthia Brewer, a US academic who completed a whole range of research into colours and colour perception. The colours suggested here are designed to ensure they can be discriminated easily, and will show up on whatever output you put your map on - e.g. colour print, screen or a digital projector (see figure 4.10, on the next page).

You can choose what type of data you have - qualitative (no order), quantitative of high/low (order), how many classes you want, and then the site will give you suggested colours. ColorBrewer is also integrated into R, so we can use exactly the same colours in the maps we generate in R (see figure 4.10, on the following page).

---

[16]https://en.wikipedia.org/wiki/The_Magical_Number_Seven,_Plus_or_Minus_Two

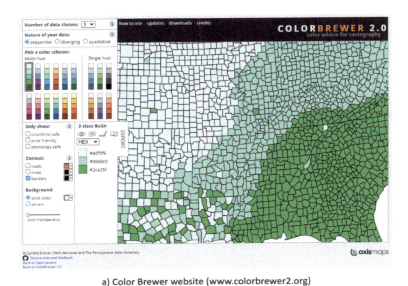

a) Color Brewer website (www.colorbrewer2.org)

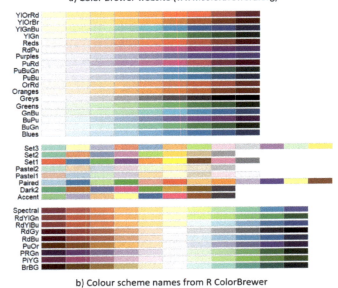

b) Colour scheme names from R ColorBrewer

Figure 4.10: a) Screenshot of the ColorBrewer website. b) Colour scheme names from R ColorBrewer.

4.6   Histograms

When deciding how to classify our data, it can be useful to look at a histogram of the data. This shows how the data is distributed, with the values (from low to high) along the x-axis (bottom) and the number of values along the y-axis (up the side).

Using our POP2105 data this is how we plot a histogram in R (figure 4.11):

```
hist(countries$POP2015)
```

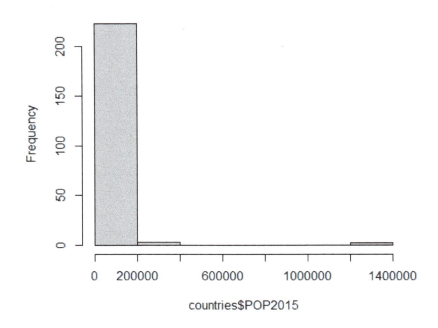

Figure 4.11: Histogram of POP2015 from countries, with the default options.

In this example, the data is quite skewed, with many more readings towards one end (the lower end in this case) and fewer readings towards the higher end (as we saw with our equal interval example). When thinking about classification methods, an equal interval would not work very well here, because most of the data points would be in one category. We can plot the histogram in R, and overlay our classifications. Have a go with this code (figure 4.12, on the facing page):

```
#select the variable
var <- countries$POP2015
#calculate the breaks
library(classInt)
breaks <- classIntervals(var, n = 6, style = "equal")
#draw histogram
hist(var)
#add breaks to histogram
abline(v = breaks$brks, col = "red")
```

See how the equal intervals don't work very well?

The code breaks$brks extract the breaks as numbers:

```
breaks$brks
[1]      0.0  229341.5  458683.0  688024.5  917366.0 1146707.5 1376049.0
```

You can see how these are shown on the histogram (figure 4.12, on the next page.

We can also extract the number of points in each category, and draw a map (figure 4.13, on page 82):

```
  tm_shape(countries) +
tm_polygons("POP2015", palette = "Greens", style = "equal", n = 6)
```

With skewed data, a Natural breaks classification is usually a good idea. See this example with Fisher (map figure 4.14, on page 82 & histogram figure 4.15, on page 83):

```
tm_shape(LSOA) +
tm_polygons("Age00to04", title = "Aged 0 to 4", palette = "Greens",
style = "jenks") +
tm_layout(legend.title.size = 0.8)
```

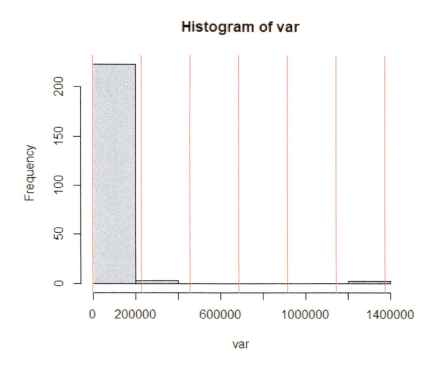

Figure 4.12: Histogram of POP2015 from countries, with equal interval classification.

```
#select the variable
var <- countries$POP2015
#calculate the breaks
library(classInt)
breaks <- classIntervals(var, n = 5, style = "jenks")
#draw histogram
hist(var)
#add breaks to histogram
abline(v = breaks$brks, col = "red")
```

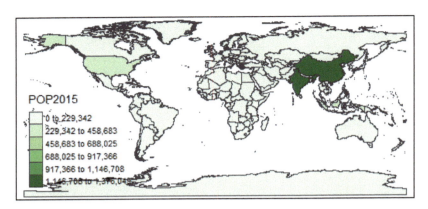

Figure 4.13: Map of POP2015 from countries, with equal interval classification.

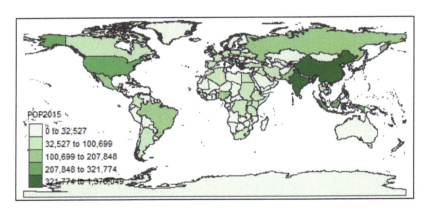

Figure 4.14: Map of POP2015 from countries, showing the 5 classes with Jenks classification.

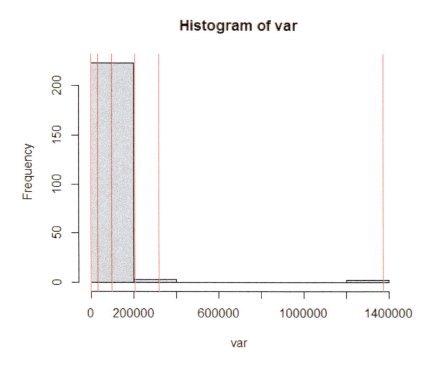

Figure 4.15: Histogram of POP2015 from countries, showing the 5 classes with Jenks classification

What's the different between Jenks and Fisher?

In most GIS programs, we talk about Natural breaks as the go-to classification method, as this usually generates a good map. R, however, likes to be different, and here we have two systems - Jenks and Fisher. Jenks and Fisher are the two academics who came up with the Natural Breaks classification system.

There are some very minor differences between how the code in fisher and the code in jenks is implemented in R. Have a look at the help file for classIntervals() for more details. Type in ?classIntervals in at the console.

4.7   Beyond Colours

We also have a range of extra bits of code we can add, to add in scale bars, north arrows and to customise the map title.

This code adds the scale bar:

```
#set scale bar
        tm_shape(countries) +
tm_polygons("POP2015", palette = "Greens", style = "jenks") +
##scale bar
tm_scale_bar(width = 0.22, position = c(0.5, 0.01))
```

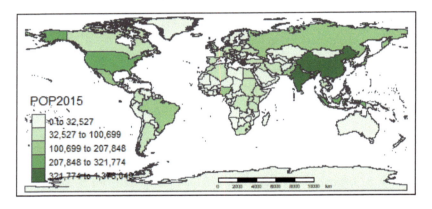

Figure 4.16: Map with scale bar added.

Note how R gives you some information in red:

Scale bar set for latitude km and will be different at the top and bottom of the map.

This is saying that the scale will vary across the map, and will be different at the top and bottom to the middle of the map. This is because we're using a geographic coordinate system rather than a projected one - more on this in Chapter 2.

You may need to adjust the position of the items on the map. The *position* parameter adjusts this. The values are between 0 and 1, with 0 being left/bottom and 1 being right/top. For example, c(0.5, 0.19) positions the scale bar 50% (0.5) from the left-hand side of the map (i.e. in the middle), and 1% (0.01) from the bottom of the map. You can also use keywords for these instead of the numbers. The keywords are top, middle, bottom, and left, center and right (note the US spelling). The width parameter specifies how wide/long the scale bar is. Try adjusting this to 0.4 and see what happens.

There are many many more options within the tmap library you can use to customise your map. For more information on the positioning, try Googling "tm_scale_bar position".

We can also use a similar approach to adding a north arrow, with the same principle applied for position.

```
tm_shape(countries) +
tm_polygons("POP2015", palette = "Greens", style = "jenks") +
##scale bar
tm_scale_bar(width = 0.22, position = c(0.5, 0.01)) +
##north arrow
tm_compass(size = 0.7, position = c(0.78, 0.03))
```

There are a range of schools of thought about whether you need a scale bar and a north arrow on a map. It really depends on who is going to be using your map, and what they're going to be using it for. These are often style considerations, and up to you whether you include these or not. Personally, I would say if the map is rotated (so up isn't north) then you must have a north arrow. Otherwise, it is down to personal preference. For scale, I would say if people looking at the map know the area and how big things are, then a scale is optional. The best suggestion is to look at what other people do and get inspiration from that.

A great book I would recommend is "Cartography: An Introduction", produced by the British Cartography Society. It is now in its second edition, and has all sorts of examples of good practice for producing maps, and simple steps on how to improve your maps[17]. The first edition is also freely available on their website[18]!

---

[17]cartography.org.uk/product-page/cartography-an-introduction-second-edition

[18]cartography.org.uk/thematic-mapping

4.8   Adding a Legend

It is always a good idea to include a key or legend in your map to explain whichever geographical phenomenon your map is showing. The default location of the legend is not great for our map, as it covers areas up. We can, however, move the legend outside the map (figure 4.17):

```
tm_shape(countries) +
tm_polygons("POP2015", palette = "Greens", style = "jenks") +
#legend position
tm_layout("Equal Interval", legend.outside = TRUE,
legend.outside.position = "bottom") +
##scale bar
tm_scale_bar(width = 0.22, position = c(0.5, 0.01)) +
##north arrow
tm_compass(size = 0.7, position = c(0.78, 0.03))
```

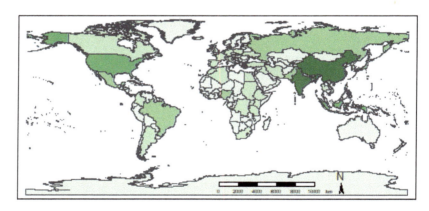

Figure 4.17: Map with legend outside the map box.

There are many many options with tmap for customizing the legend. For more details, have a look at:

- rdocumentation.org/packages/tmap/versions/3.3-3/topics/tm_ fill Search for *legend.format* and look at the options there
- rdocumentation.org/packages/tmap/versions/3.3-3/topics/tm_ layout has some useful info on legend position and size (search for *legend.*)
- jla-data.net/2017/09/20/2017-09-19-tmap-legend shows you how to incorporate custom symbols next to the values (e.g. £ or $ for currency, % for percentages).

4.9 Dynamic Maps

The tmap library has two different viewing options - plot mode which is what we'ave been using so far, and view mode which provides a basemap and the ability to zoom in and out (figure 4.18, on the following page). Try this code:

```
#set tmap to view mode
tmap_mode("view")
#plot using qtm
qtm(countries)
#plot using tm_shape
tm_shape(countries) +
tm_polygons("POP2015", palette = "Greens", style = "jenks")
```

tmap creates an interactive map - so you can zoom in, and click on different polygons. This can be really useful for exploring and investigating data. Try some of the different options.

When you have finished, remember to change the tmap_mode back to plot. Otherwise weird things can happen:

```
#return tmap to plot mode
tmap_mode("plot")
```

4.10 tmap v4

When you install tmap in R, you get version 3.3-4. Currently a version 4 is being developed and is not that far away from being released. A very good overview is available at mtennekes.github.io/tmap4 which I would recommend you take a look at. There are quite a few changes

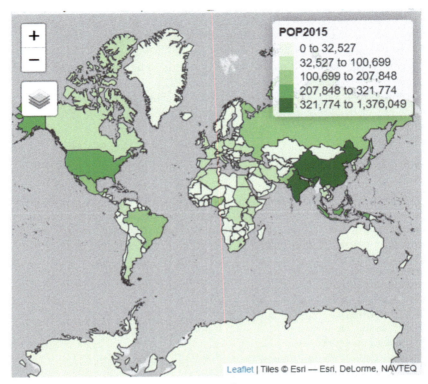

Figure 4.18: Map using the tmap plot mode.

which improves the tmap library. I'm not sure when the new version will be available and will install by default in R. At the moment you can preview the new version by installing it:

```
install.packages("remotes")
remotes::install_github('r-tmap/tmap')
library(tmap)
```

If you want to go back to version 3, you will need to uninstall the new library, through the packages tab in RStudio. If you just want to try it out, I would recommend setting up an account on Posit.cloud where you can install it and try it out without changing your main R setup.

5. Raster Data

5.1 Learning Outcomes

After reading this chapter, you will:

- Understand what raster data is.
- Know typical applications of raster data and what types of data it is often used to represent.
- Be aware of different resolutions of raster data.
- Understand that satellite data is a type of raster data.
- Understand different wavelengths of satellite data.
- Be able to work with raster data in R using (libraries).

Building on creating maps with vector data (from Chapter 4), this chapter moves on to working with raster data and plotting raster maps. We will use some of the same tools and principles to plot maps with raster data, but there are also a few key differences. We will also talk about working with raster and vector data together.

5.2 Raster Data Structure

We have already discussed a bit about the difference between raster data and vector data in Chapter 2. Remember we have difference conceptual models for vector data and raster data: vector data is a series of discrete objects, (sometimes) with gaps in between, raster data is a continuous grid (see figures 5.1 and 5.2 on the following page):

Raster data is most commonly used for continuous variables, i.e. data that has a value at every location. A typical example might be elevation - everywhere has an elevation value. Raster data is structured as a grid over a set area. The grid is typically made up of *square* cells, and each cell will have a numeric value which represents the data. The cells are a specific size, which is the resolution of the raster data; resolutions can very from relatively low detail e.g. 1km, to very high detail, e.g. 15cm.

a)   b)   c)

Figure 5.1: The three different types of vector data, a) points, b) lines and c) polygons. Note how these are all discrete objects.

a)

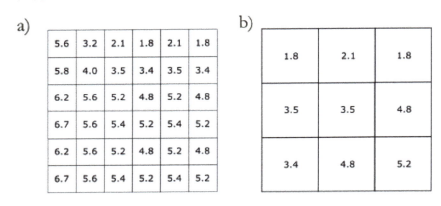

b)

c)   d)

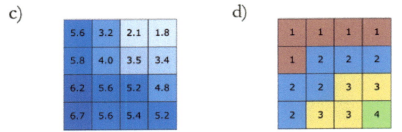

Figure 5.2: Examples of raster data. The resolution is higher in a) and lower in b). Raster data can be used for c) quantitive data and d) qualitative data. Note that this is continuous data - every location (within the raster coverage) has a value.

> **Squares, rectangles, hexagons and variable squares**
>
> Traditionally, raster cells are square, but they don't have to be. As long as they tessellate to fill the area the data covers, they can be any shape. Rectangles are used sometimes, as are triangles. Hexagons have become very common recently, with the H3 system designed by Uber (uber.com/blog/h3) a really useful example.
>
> A TIN is a variation on this concept of a regular grid. TIN stands for Triangulated Irregular Network and is made up of varying sizes of triangles. This allows the resolution of the data to vary over space, to show some areas in high detail when required, and some areas in lower detail to save on file space.
>
> The ONS (Office for National Statistics, UK) is planning on releasing some of the 2021 Census data in a varying resolution square grid. This means that Census data will be in a regular grid which makes performing comparisons over time easier, and also allows some areas to have 0 values, which is not currently possible with the Output Area approach. The varying resolution means that there will be difference cell sizes, and this is to allow preventing disclosure (being able to identify specific individuals), with more detail in urban areas and less in rural areas.
>
> The ONS variable grid is usually released as what we call vector grids. This is a vector polygon format (rather than raster) and is used because there is no standard way of encoding these types of shapes in raster data. Similarly, the H3 hexagons are a standardised system, but are usually released as vector grids to allow any GIS software to work with them.

Raster data is very common in the environmental sciences. Typical examples of raster data include elevation, mean temperature and rainfall. These datasets use continuous values to represent the data. You also get some social science examples, particularly population. For example, the number of people per $1\,km^2$, per $100\,m^2$ or whatever the resolution is (see figure 5.3, on the next page).

Categorical data is also very common in raster data, for example in land cover. Integers are used to represent different land cover types, for example 1 = urban, 2 = coniferous forest, 3 = deciduous forest etc. (see figure 5.4, on page 93).

Finally, satellite and aerial photography data are also classified as raster data. They are made up of individual pixels, much like a photo taken

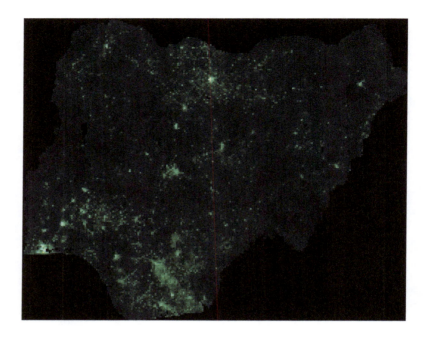

Figure 5.3: Population for Nigeria shown at 1km resolution.

on a mobile phone. Each pixel, or cell, stores three separate numbers which make up a colour which is shown on screen when the raster data is loaded (see figure 5.5, on page 94).

Raster data like the examples above (elevation, population, etc.) just have one set of values and are what we call single band data. Showing these on a map is fairly straight forward, and you can show these using shades of one colour, or a diverging set of colours (e.g. red-blue) depending on what you are showing.

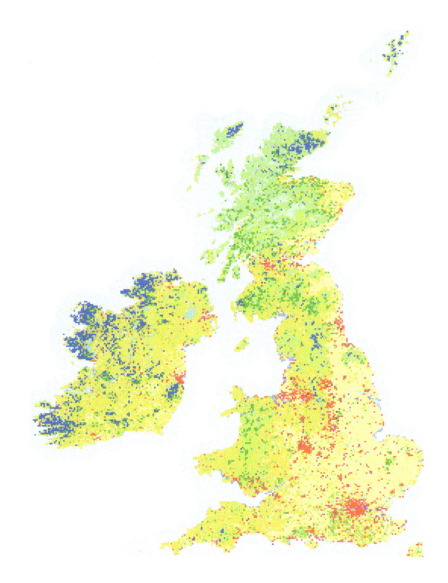

Figure 5.4: Categorical raster data, showing land cover for the UK.

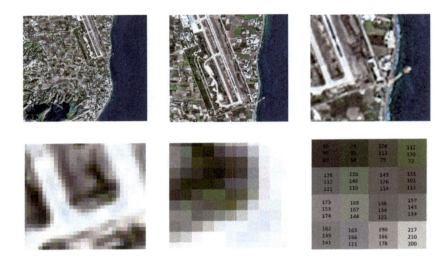

Figure 5.5: A Sentinel satellite image from Santorini, Greece. Notice that as you zoom in the image gets less detailed, until you can see the individual pixels. The numbers in the pixels represent the colours, showing the amount of red, green and blue (out of 255) of that pixel.

5.3   Satellite Data and Aerial Photography

Satellite data and aerial photography can be a little more complex to show, because they have more data in them. Computers use three different colours to show information on a computer monitor - Red, Green and Blue. These colours are known as additive primaries, because when you add the three colours together (Red, Green and Blue) you make white. This works very differently to printing colours on paper, which are known as subtractive. These are Cyan, Magenta, Yellow and blacK, and is known as CMYK. We have to remember this when printing maps on paper, although most GIS programs handle this automatically. For more info, have a look at BCS Cartography: An Introduction, p30-31 [5].

This means that for a satellite image, there are three different parts that are needs to show the data. These are called bands, with a red band, a green band and a blue band (matching the three colours mentioned earlier). For all aerial photography and some satellite images, these have data on what is visible in the red, green and blue wavelengths. When shown through the red, green and blue bands, we create an image that looks like we would see it in reality - this is sometimes called a true colour composite (see figure 5.6):

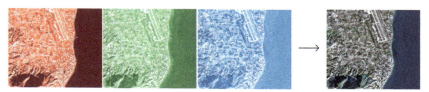

Figure 5.6: The individual bands of the RGB image combined to make a true colour composite. A Sentinel satellite image from Santorini, Greece.

This is how normal images are stored on a computer, and how most aerial photography (taken from a UAV or a plane) and some satellite imagery (taken from a satellite in space) work. These all fall in the 400 to 700 nanometre wavelength section. However, visible light is just a small part of the electromagnetic spectrum (see figure 5.7). Most satellites (and some planes and UAVs) can record data from outside the visible wavelengths - i.e. beyond what we can see. Infrared is very common, and radar is also sometimes used. Shorter wavelengths (ultraviolet, X-rays etc.) are not normally used because they can be (in quantity) harmful to human health.

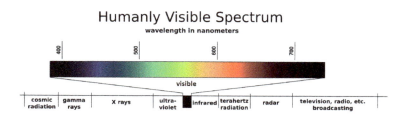

Figure 5.7: The electromagnetic spectrum, with the visible wavelengths highlighted.

Visible light is a part of the light spectrum, but there are many other wavelengths of light on the spectrum as well. Sensors on satellites can capture some of these. Infrared is a very common wavelength to capture and has lots of applications in remote sensing. The wavelength is how we define the electromagnetic signal the satellite receives - there is a wide spectrum, inducing infrared, ultraviolet, X-rays and many others. Not all are usable for remote sensing - using X-rays and UV have health implications! Infrared is widely used, and can provide really useful information on plant health, as well as how much heat is being emitted by objects.

Red, green and blue appear on the spectrum as visible light. If a sensor collects images in these wavelengths, then these can be combined to create an image that looks more or less like we would see in reality. This is called a true colour composite (as shown in figure 5.6, on the preceding page). However, we don't have to show the red wavelengths as red, green wavelengths as green or blue wavelengths as blue on the computer. We can show different channels. A very common combination is showing infrared as red, red wavelengths as green and green wavelengths as blue. This is really useful because healthy vegetation reflects infrared very strongly, whereas unhealthy vegetation does not. So in our output, healthy vegetation appears as bright red, and unhealthy as dark red.

Infra-red data is sometimes used to create an NVDI - Normalised Vegetation Difference Index (as shown in figure 5.8, on the next page). This is a measure of how much vegetation there is, and how healthy it is. This just uses the red and infrared bands, and creates a greyscale image - with high values representing healthy vegetation and low values representing unhealthy vegetation [7].

Another well-known raster product is LiDAR. This brings us to another important point - active and passive sensors. Passive sensors work to collect EM waves that are reflected by objects. For example, the sun shines, reflects off the ground and this is detected by the sensor. This is how the majority of sensors work, including everything we have discussed so far. We have a great energy source (the Sun) that emits a whole range of wavelengths of light which then reflects off objects.

The other type of sensors are active sensors. Rather than relying on the Sun, these sensors emit their own wavelengths of EM radiation. LIDAR is a very common example - it emits energy, which is then reflected and detected. This works well when the Sun isn't available - e.g. indoors - and can create a very detailed set of data. It also doesn't have to be

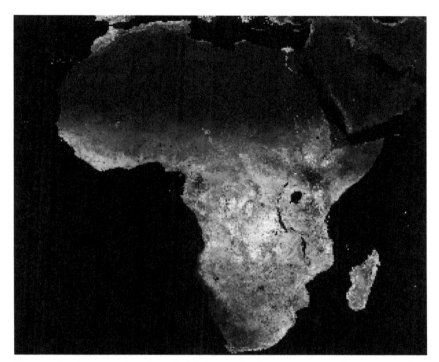

Figure 5.8: An example of an NDVI (Normalised Vegetation Difference Index) image, calculated using red and near infrared wavelengths.

'top-down' - you can use LIDAR to capture 3D objects.

All satellite and aerial data need some processing from when they're captured to being available in a GIS. This involves a range of tasks, including correction of the values (because the atmosphere can distort the light received) and georeferencing - giving the data coordinates. Most of the raster data you will come across has been processed already, so we don't need to worry about this, but if you're interested see [6] for more details.

5.4 Working with Raster data in R

However, the data is collected, we end up with some raster data that we can then work with in a GIS. Like vector data, it needs a coordinate system specified so that the GIS knows where in the world it is.

The next chunk of code will show you how to read raster data into R, and to visualise it.

We are going to work with the terra package for raster data.

From raster to terra (and stars)

Previously we had the raster package, but this is being replaced over time. Much like how sf is replacing sp for vector data, raster data is having a similar change. There are actually two new packages worth mentioning - terra and stars.

terra is more the direct replacement for the raster package. It is significantly faster than raster for many operations and has more long-term support. We will use terra in this practical and is the one I would recommend for new users. Most raster functions have equivalents in terra and it works is more or less the same way. There is a nice write-up of the differences between the two at oceanhealthindex.org/news/raster_to_terra.

One annoyance is that the terra package will not directly work with sf data types. There are convertor functions to convert from sf to SpatVector (which is the vector format terra uses) so it is not a major problem, it just makes the code a bit more clunky. Some code will work with sf data types, but it is a bit hit-and-miss, so you are probably going to need to do some conversion at some point. There is some interesting discussion at github.com/rspatial/terra/issues/89.

The other package worth mentioning is stars. This performs a similar function to terra but is a bit more complex to use. There are some good guides available, and stars can deal with more complex data types that terra. One particular example is it handles a variety of grid formats, including curvilinear grids, which are very common in climate science, particularly NetCDF files. See r-spatial.github.io/stars/articles/stars4.html for more details of this, including examples.

Install the package if you need to:

```
install.packages("terra")
```

We also need some raster data to work with. Go to: worldpop.org/geodata/summary?id=32968 and download the nga_ppp_ 2020_1km_Aggregated.tif file and save it in your working directory.

Use this code to read in the data:

```
#load library
library(terra)

# read in raster data
nigeriaPop <- rast("nga_ppp_2020_1km_Aggregated.tif")
```

Now the raster data is loaded into R as a variable called nigeriaPop. We can view some information about the raster data:

```
# display summary
nigeriaPop
# Our data has one band (nlyr)

class       : SpatRaster
dimensions  : 1156, 1441, 1  (nrow, ncol, nlyr)
resolution  : 0.008333333, 0.008333333  (x, y)
extent      : 2.67375, 14.68208, 4.26625, 13.89958  (xmin, xmax...
coord. ref. : lon/lat WGS 84 (EPSG:4326)
source      : nga_ppp_2020_1km_Aggregated.tif
name        : nga_ppp_2020_1km_Aggregated
min value   :                   0.193234
max value   :                 93660.921875

# resolution (in the CRS)
res(nigeriaPop)
[1] 0.008333333 0.008333333
# our data is in WGS84 (Lat/Long) so 0.0083... degrees = 1km

# what CRS is it using?
crs(nigeriaPop)
```

There is a lot of text in the crs() output but the key bits are WGS 84 near the beginning and EPSG, 4326 near the end.

We can plot the raster data using base R (see figure 5.9):

```
plot(nigeriaPop)
# and add a title
plot(nigeriaPop, main = "Nigeria Population 2020")
```

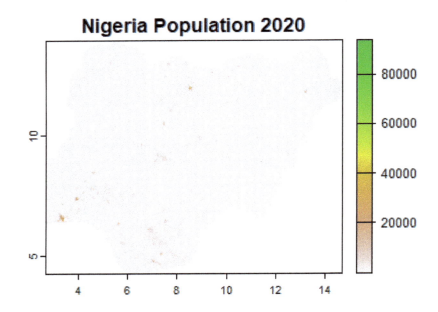

Figure 5.9: Basic map of the Nigeria Population data from 2020, with a title.

Our basic plot is ok, but not very exciting or pretty. We can improve this a bit.

By default, the data doesn't show very well because we have urban areas with a very large population, and rural areas with a very low population. If we do a histogram, we can see that we have a lot of very low values. If we log the data, it looks a bit better (see figure 5.10, on the next page).

```
# histogram
hist(nigeriaPop)
# log
plot(log10(nigeriaPop))
```

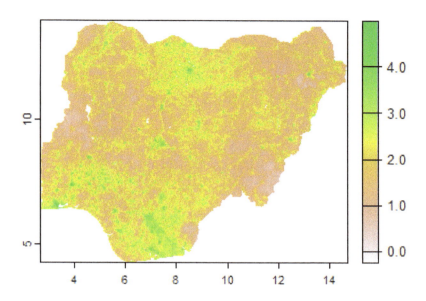

Figure 5.10: Map of the Nigeria Population data from 2020, with a log10 transformation.

We can also improve this a bit more by customising the colours (see figure 5.11, on the following page):

```
plot(log10(nigeriaPop), colNA="black", col=colorRampPalette(c("#121212",
"#123620", "#108650", "#80D6C0", "#DFDF3B"),0.5)(110))
```

The tmap library is also useful:

```
#load library
library(tmap)
#qtm
qtm(nigeriaPop)
#qtm + log
qtm(log10(nigeriaPop))
#tm_shape
tm_shape(nigeriaPop) +
  tm_raster("nga_ppp_2020_1km_Aggregated.tif")
```

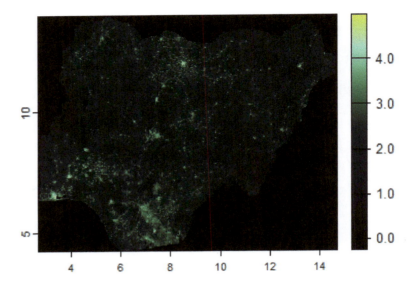

Figure 5.11: Map of the Nigeria Population data from 2020, with a log10 transformation and improved colours.

```
tm_shape(nigeriaPop) +
  tm_raster("nga_ppp_2020_1km_Aggregated.tif", palette = "Greens")
```

The default style (Pretty) isn't that good. Fisher is much better, but is quite slow to calculate. log10_pretty is a bit better, and very quick to calculate (see figure 5.12, on the next page).

```
tm_shape(nigeriaPop) + tm_raster(style = "fisher")
# fisher is quite computationally heavy

tm_shape(nigeriaPop) + tm_raster(style = "log10_pretty")
# log10 pretty is quicker to calculate
```

5.5   Categorical Data

As well as quantitative data, we can also work with categorical raster data. This is data that uses integers to represent the data, and each

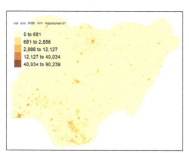 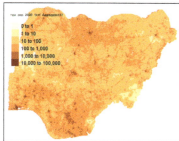

Figure 5.12: Map of the Nigeria Population data from 2020, Fisher classification
(left), Log10 Pretty classification (right).

value represents a different category. For example, Corine is a very
common land cover data set that covers Europe. Here, different colours
are used to show different categories (see figure 5.13, on the following
page).

```
#read in the corine data
corine <- rast("corine/U2006_CLC2000_V2020_20u1.tif")
#plot data
plot(corine)
```

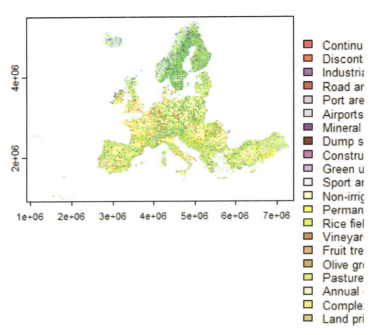

Figure 5.13: The Corine land cover data for Europe. Note how different colours represent different land cover types. In the raster data behind the scenes, each colour has a specific code. Of course the layout of the legend could be improved here!

> **Corine Land Cover (CLC)**
>
> Corine is a very commonly used land cover data set. It is collated by the European Environment Agency, and covers the whole of Europe. It is available in both raster and vector format and also shows change over time, with data available for 1990, 2000, 2006, 2012 and 2018.
>
> The Corine Land Cover (CLC) is a European programme, coordinated by the European Environment Agency (EEA), providing consistent information on land cover and land cover changes across Europe. CLC products are based on (visual or semi-automated) interpretation of high-resolution multispectral satellite imagery by the national teams of the participating countries - the EEA member or cooperating countries.
>
> More details and the data are available from land.copernicus.eu[a]. To download the data you need to register, which is a bit of a fiddly process so I have included it in the data you download from the book website.
>
> ---
>
> [a]https://land.copernicus.eu/en/products/corine-land-cover/clc-2000?hash=7be11e321982cd1df8a9e71b7ebf6e7839fd1424

5.6 Cropping Data

Sometimes you will need to subset the data spatially - select data for a certain geographic location. For example, from the Corine data for the whole of Europe, we might only be interested in one or two countries. We will cover some general methods for this including sub-setting data using [,] in Chapter 10 - Spatial Data Wrangling. However, for raster data we can't use this technique for sub-setting because raster data doesn't have an attribute table in the same way vector data does. Instead, we have to use spatial data to select the area we're interested in.

The code over the next few pages will show how we can use the country outline for UK to crop out the UK-based Corine data. This is a very common approach when working with raster data, and can be a good way to reduce the file size of the data you're working with, and therefore improve processing times.

```
library(sf)
#read in world countries
countries <- st_read("world_countries.shp")
```

```
#plot data
qtm(countries)
#view head information
head(countries)
```

```
Simple feature collection with 6 features and 11 fields
Geometry type: MULTIPOLYGON
Dimension:     XY
Bounding box:  xmin: -61.89111 ymin: -18.01639 xmax: 50.37499 ymax: 42.6619
Geodetic CRS:  WGS 84
  FIPS ISO2 ISO3 UN               NAME    AREA POP2005 LIFEEXP LIFEEXPM
1   AC   AG  ATG 28 Antigua and Barbuda     44      83   75.00    72.56
2   AG   DZ  DZA 12              Algeria 238174   33268   73.06    71.02
3   AJ   AZ  AZE 31           Azerbaijan   8260    8563   70.12    67.09
4   AL   AL  ALB  8              Albania   2740    3082   76.30    73.42
5   AM   AM  ARM 51              Armenia   2820    3015   74.00    70.58
6   AO   AO  AGO 24               Angola 124670   17913   49.63    48.21
  LIFEEXPF  INFMRT                       geometry
1    77.36  10.028 MULTIPOLYGON (((-61.73806 1...
2    75.22  34.176 MULTIPOLYGON (((2.96361 36....
3    73.14  41.062 MULTIPOLYGON (((45.31998 39...
4    79.72  16.108 MULTIPOLYGON (((19.43621 41...
5    77.30  21.009 MULTIPOLYGON (((45.21137 41...
6    51.04 104.349 MULTIPOLYGON (((11.775 -16....
```

The head() function shows us the attribute data from countries. We are particularly interested in the NAME column because this will allow us to select out the UK.

```
#show entries in NAME column
countries$NAME
#find which one is UK
which(countries$NAME == "United Kingdom")
```

```
[1] 197
```

The which() function is very useful because it allows us to select a specific item, in this case the one which contains the UK. R works a lot with row and column numbers - so the output from this: 197 tells us that the UK is the 197th row in our data set.

It's always checking that R has got the right row, so let's check this now. head() will show us the attribute data for this row, and qtm() will draw us a map of that row:

```
head(countries[197,])
qtm(countries[197,])
```

If these output the UK, this confirms that 197 is the right number. If they don't, then we have picked the wrong value.

We can use the crop() function to cut out a section:

```
tmp <- crop(corine, countries[197,])
```

This command will take the UK outline (countries[197,]) and use it to crop out the related area from the Corine data set. However when we run it, we get an error:

```
Error: [crop] extents do not overlap
```

This is saying the two spatial datasets don't overlap. However, we know this isn't the case because the UK is within the European area of the Corine data. So we're going to take a guess that the CRS (coordinate reference systems) of the two data sets are not the same. Some functions will tell us this, but the crop function doesn't.

We can check this by using the crs() function:

```
crs(corine)
crs(countries[197,])
```

There is a lot of text in this output, but the key bits are in the first line.

```
crs(corine)
[1] "PROJCRS[\"ETRS89-extended / LAEA Europe\",
      BASEGEOGCRS[\"ETRS89 crs(countries[197,])
[1] "GEOGCRS[\"WGS 84\", DATUM[\"World Geodetic System 1984\",
```

The corine dataset is in ETRS89, also known as LAEA Europe. The countries data is in WGS 84, what we also know as latitude longitude. This confirms the CRS are different which would cause the error we've seen. We can reproject one layer to a different coordinate system.

In many cases it doesn't matter which layer we reproject, but if we're working with more than two datasets, it's good practice to reduce the number of reprojections wherever possible. For example, if you had two layers in British National Grid (BNG) and one layer in WGS 84, you would reproject the WGS 84 layer to BNG.

There is an exception to this: if the areas your data set cover are different. For example, we couldn't reproject our Corine data to BNG because it covers the whole of Europe and BNG only covers Great Britain.

In our case, we are going to reproject the UK data to the ETRS89 (the CRS of the corine layer).

```
#save UK as a new variable
UK <- countries[197,]
#reproject
UK_new <- project(UK, crs(corine))
```

This code takes the CRS of corine and applies it to the UK. However we get another error:

```
Error in (function (classes, fdef, mtable)   :
unable to find an inherited method for function 'project' for
signature '"sf"'
```

This means that R can't find a method within the function project() to work with data in the sf format.

We can look at the help for project() by running ?project and see that it says x (out input) has to be a SpatRaster or SpatVector. Our x (UK) is currently in the sf format, so we need to convert it (see box on **From raster to terra (and stars)** earlier for more details):

```
#convert sf to SpatVector
UK_spatVect <- vect(UK)
#reproject (this time without errors!)
UK_new <- project(UK_spatVect, crs(corine))
#crop the data (again without errors!)
tmp <- crop(corine, UK_new)
#plot the map to check output
plot(tmp)
```

So you've now cut out the UK from the Corine dataset. We can also use the same approach to use spatial data to crop out a selected area. The process is much the same, but rather than creating a subset, you use a variable you have read in from a file. They still both need to have the same coordinate system, so you are likely going to need to reproject the layer.

We have also encountered a series of error messages. This is usual when working in R, so when you encounter an error message don't be put off! Hopefully you can see how I have worked through them.

Different people work in different ways, but sometimes I find it hard to see what is going on at each stage of the R code. Tools like QGIS provide a visual output at each stage, so it is much easier to see what is going on. However, it is much more difficult to document and replicate steps in QGIS, so there are pros and cons to each approach. Try different ones and do what works best for you!

6. Interactive Maps

6.1 Learning Outcomes

After reading this chapter, you will:

- Be aware that we can use interaction and animation to show a range of spatial data including temporal (time) data.
- Know how to create interactive maps using the tmap library.
- Be aware of the Shiny service from Posit and how to create and publish interactive Shiny apps.
- Understand methods for creating animated maps in R & QGIS.

All the maps you've created so far are static, i.e. they're fixed images that can be shown on screen, included in a report or printed out. This is by far the most common type of map and what you will likely be making 90% of the time. However, there are a few other options that are worth mentioning.

6.2 Interactive Maps

The main one here is interactive maps, or what are sometimes called slippy web maps. This is when you can load up the map on a web page, and scroll around and zoom in or out. There are a range of tools you can use to create these. The tmap library has one built in - tmap's view mode.

We can use exactly the same code we used in chapter 4 to create the map. Then we have a command to change the plot mode to "View" which is the interactive map mode (see figure 6.1, on the next page).

```
tmap_mode("view")
qtm(countries, fill="POP2015")
```

This gives us a basic interactive map. There are many ways you can

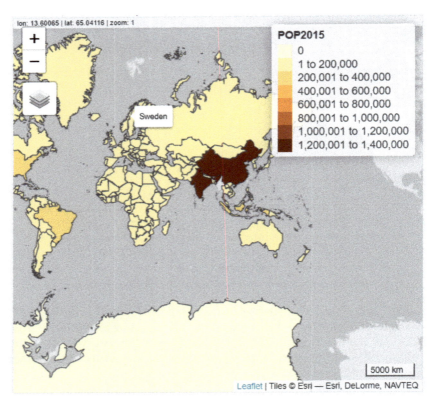

Figure 6.1: A basic interactive map from tmap. You can zoom in and out, drag the map around and click on different countries.

extend this - initially using the tm_shape code instead (see figure 6.2, on the facing page):

```
tm_shape(countries) +
tm_polygons("POP2015", palette = "Greens", style = "jenks")
```

These can also be saved as HTML files, which can then be hosted online and shared with others.

Behind the scenes, tmap adds a JavaScript library to manage and run the interactive map. If you upload this to a website, it will work and users will be able to interact with it. However, this is what is called client

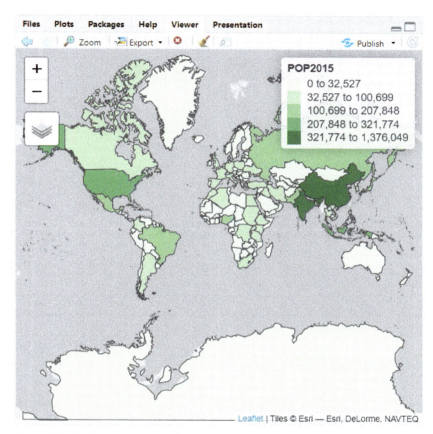

Figure 6.2: An interactive map from tmap using exttttm shape.

side which means that the user's web browser does all the processing to render the map. This is fine with a small data set, but if you want to share a large data set, then a different approach is needed. See r-tmap.github.io for more details.[19]

---

[19]https://r-tmap.github.io/tmap/reference/tm_view.html

6.3 Shiny

Shiny is an R library that allows you to build interactive web apps right from an R script.[20] They can do many different things, including mapping.[21]

Shiny makes it easier to share web map apps, and you can easily share them with the public on shinyapps.io, Posit's hosted service. These are free for certain usage and then there are a series of paid levels if your app becomes really popular.

6.4   Other Mapping JavaScript Libraries

However, both tmap and Shiny have limited flexibility in terms of how you lay out your app, and how you customise it. If you want more flexibility, you can use a JavaScript mapping library directly. These are more complex to setup, but give you more control. There are a number of completely free and open source ones you have to host yourself (Leaflet, OpenLayers), and there are ones you can pay for with higher usage and for support (Mapbox, Google Maps).

6.5   Animations

Currently, we've only discussed maps where the user interacts directly with the map. The other option is animation, often used to show temporal data. Animations are a really powerful visualisation tool which can really aid data presentation. They can also make large datasets completely unintelligible, so the tools need to be used with caution. A very common approach is to create a series of still maps in our GIS and then use an animation tool to "paste" all the images together to create an animation.

QGIS has some very useful animation tools, using the Temporal Controller. Interestingly, this started life as a QGIS plugin (the Time Manager plugin) and it was so useful it was subsequently incorporated into

---

[20]Overview   of   Shiny: `https://shiny.posit.co/r/getstarted/shiny-basics/lesson1/`

[21]Blog post for creating a shiny web map: `https://rviews.rstudio.com/2019/10/09/building-interactive-world-maps-in-shiny/`

the core QGIS program. There are a number of good tutorials available for this.[22][23]

There are a number of approaches to animating maps in R. You can do this in tmap.[24] Other approaches make use of ggplot and gganimate. We've not focused on ggplot in this book, but it uses a similar approach to tmap so the skills you already have will help you learn this. Do have a look at this tutorial[25] and this tutorial[26] to get an idea about creating animated maps in R.

---

[22]https://www.qgistutorials.com/en/docs/3/animating_time_series.html

[23]https://www.youtube.com/watch?v=PnDiG2h6N8M

[24]Check out https://rdrr.io/cran/tmap/man/tmap_animation.html for an example tutorial

[25]https://aberdeenstudygroup.github.io/studyGroup/lessons/Visualising%20Spatial%20Data/Visualising_Spatial_Data_in_R/

[26]https://geospatialtraining.com/creating-animated-maps-with-r/

7. Spatial Libraries - sf and sp

7.1 Learning Outcomes

After reading this chapter, you will:

- Understand that sf is one of two major spatial libraries in R.
- Be aware of the sp library and that is it different to sf.
- Be able to identify sp code in R.

We've already talked about different libraries in R, and the fact that they add new functions and tools (see Chapter 3). We've looked a bit at different spatial libraries - sf, tmap, terra/stars and a few other ones. However, there is another one which we haven't looked at yet: sp.

So far we've used sf to work with spatial data. sf is, in fact a replacement for the sp library, with this replacement taking place from around 2020 to 2023. At the time of writing (2024) sp is no longer available, but you will still probably come across sp code on the internet, so it is worthwhile spending a little bit of time introducing the two libraries and explaining the differences between them.

7.2 Key Differences

sp was the original spatial library developed for R. It was first released in 2005[27] and was a great development because for the first time it allowed R to handle spatial data. It has developed over the years since, and has a number of limitations, including some which came to light with the advance of data science and the tidyverse multiverse (see later). The sf library was developed to address some of these issues, and it is a much more recent library - it was first released in October 2016.

---

[27]https://cran.r-project.org/doc/Rnews/Rnews_2005-2.pdf

In the first few years of usage (2016-2020) sf took a while to get used -
most people still made use of the sp library. As data science advanced,
and as the capabilities of sf developed, more and more people moved
over. Until about 2021, there were some analysis that could only be
done using the sp library. After 2021, this was no longer the case and
any spatial analysis that can be done in sp can also be done in sf (plus
some more types which we will get to). Now sp has been officially
depreciated at the end of 2023.

The key difference between the two is how they handle spatial data. sp
uses what R calls S4 data types. This is to do with the structure of how
R stores data, specifically spatial data and the coordinates for the spatial data. sp stores the spatial data in a separate part of the variables.
One of the major improvements provided by sf on this is that it stores
the spatial element of the data within a normal (S3) data type. The coordinate information is stored within a geometry column. This means
that spatial data is stored in a standard data frame, which means that
many different operations, including sub-setting, can be applied to sf
data. This also means that many tools from the tidyverse, including
dplyr library, which is very popular in data science, and be easily applied to spatial data. This was not the case when using the sp library.
Everything was possible, but it was just rather a lot more complicated!

Let's have a look at an example.

```
countries <- st_read("world_countries.shp")

head(countries)

Simple feature collection with 6 features and 11 fields
Geometry type: MULTIPOLYGON
Dimension:     XY
Bounding box:  xmin: -61.89111 ymin: -18.01639 xmax: 50.37499 ymax: 42.6619
Geodetic CRS:  WGS 84
  FIPS ISO2 ISO3 UN               NAME   AREA POP2005 LIFEEXP LIFEEXPM
1   AC   AG  ATG 28 Antigua and Barbuda     44      83   75.00    72.56
2   AG   DZ  DZA 12             Algeria 238174   33268   73.06    71.02
3   AJ   AZ  AZE 31          Azerbaijan   8260    8563   70.12    67.09
4   AL   AL  ALB  8             Albania   2740    3082   76.30    73.42
5   AM   AM  ARM 51             Armenia   2820    3015   74.00    70.58
6   AO   AO  AGO 24              Angola 124670   17913   49.63    48.21
  LIFEEXPF  INFMRT                         geometry
1    77.36  10.028 MULTIPOLYGON (((-61.73806 1...
```

```
2    75.22  34.176 MULTIPOLYGON (((2.96361 36....
3    73.14  41.062 MULTIPOLYGON (((45.31998 39...
4    79.72  16.108 MULTIPOLYGON (((19.43621 41...
5    77.30  21.009 MULTIPOLYGON (((45.21137 41...
6    51.04 104.349 MULTIPOLYGON (((11.775 -16....
```

This is the data we've been working with already. Note where the co-ordinates are stored (in the geometry column) and the CRS information (in the header above the rows of data).

To read in data in sp format, we need some new libraries, and a different line of code.

> Just to note, both bits of code are reading in from the same shape file. Both sf and sp can read shapefiles (and many other spatial file types).

> Also, we can name variables whatever we like. I have used _sp and _sf to identify each copy of the St Helens data in this example, but this is just a naming convention. Just because a variable is called data_sf doesn't necessarily mean it is in sf format - you can always use class(data_sf) to check.

> It is also worth mentioning that this code may not work for you. In early 2024 the sp library was retired and you won't be able to install it in R directly from CRAN (using install.packages("sp"). You can still install it to be able to reproduce old code, but it is a more complex process. All the output will be included below so you can always have this as a reference.

```
library(sp)
library(rgdal)
#note the warning about rgdal being retired during 2023 when you load
this library

countries <- readOGR("world_countries.shp")
```

sp data is structured differently. Note in the environment tab is lists countries_sp as Large SpatialPolygonsDataframe. If we run class() on them both we can see the difference:

```
> class(countries_sf)
[1] "sf"            "data.frame"

> class(countries_sp)
[1] "SpatialPolygonsDataFrame"
attr(,"package")
[1] "sp"
```

To get the attribute table, we need to use this:

```
head(countries_sp@data)
```

	FIPS	ISO2	ISO3	UN	NAME	AREA	POP2005
0	AC	AG	ATG	28	Antigua and Barbuda	44	83
1	AG	DZ	DZA	12	Algeria	238174	33268
2	AJ	AZ	AZE	31	Azerbaijan	8260	8563
3	AL	AL	ALB	8	Albania	2740	3082
4	AM	AM	ARM	51	Armenia	2820	3015
5	AO	AO	AGO	24	Angola	124670	17913

	LIFEEXP	LIFEEXPM	LIFEEXPF	INFMRT
0	75.00	72.56	77.36	10.028
1	73.06	71.02	75.22	34.176
2	70.12	67.09	73.14	41.062
3	76.30	73.42	79.72	16.108
4	74.00	70.58	77.30	21.009
5	49.63	48.21	51.04	104.349

Note the @ sign - this is one of the key practical differences. sp uses different slots to store information. You can run slotNames() to get a list:

```
slotNames(countries_sp)
[1] "data"      "polygons"    "plotOrder"   "bbox"     "proj4string"
```

Each of the slots contains information, some geographic and some not geographic.

- @data contains the attribute table, as we saw earlier.
- @polygons contains the coordinates of each of the polygons - similar to the geometry column in *sf* format.
- @plotOrder contains an integer listing the order of plotting the polygons. This is only important if polygons overlap (unusual in

one layer of GIS data) or multiple layers are plotted at the same time (very common in GIS).

- @bbox is the bounding box, a summary of the geographic location of the data:

```
> countries_sp@bbox
   min     max
x -180 180.0000
y  -90  83.6236
```

- @proj4string is the projection and coordinate system information

```
> countries_sp@proj4string
Coordinate Reference System:
Deprecated Proj.4 representation: +proj=longlat +datum=WGS84 +no_defs
WKT2 2019 representation:
GEOGCRS["WGS 84",
    DATUM["World Geodetic System 1984",
        ELLIPSOID["WGS 84",6378137,298.257223563,
            LENGTHUNIT["metre",1]]],
    PRIMEM["Greenwich",0,
        ANGLEUNIT["degree",0.0174532925199433]],
    CS[ellipsoidal,2],
        AXIS["latitude",north,
            ORDER[1],
            ANGLEUNIT["degree",0.0174532925199433]],
        AXIS["longitude",east,
            ORDER[2],
            ANGLEUNIT["degree",0.0174532925199433]],
    ID["EPSG",4326]]
```

One important thing to remember is not to do head(countries_sp) on a sp object. This will print out the first 6 rows of each slot - and the first 6 rows of the polygons slot contains a lot of coordinates! If you do run this by accident, don't worry. R will eventually stop printing material - but it might take a little while.

The sp format bolted on the spatial data to an existing data frame in a way that meant that the dataframe couldn't be edited very easily. The two elements (@data and @polygons) were very closely related but not formally linked. The first row in @polygons relates to the first row in @data - but you could easily edit one and not the other, which would mean the rows didn't line up any more - which could case all sorts of problems!

The sf format solved this problem by integrating the spatial (@polygons) and non-spatial (@data) table back into the dataframe, so you only have one thing to edit. If you run class() on a sf data frame, you can see it says it is both a data.frame and sf (simple features) object:

```
> class(countries_sf)
[1] "sf"         "data.frame"
```

7.3   Plotting Maps: A Comparison

When using the tmap library, we are quite lucky because tmap will plot both sf and sp data. The code and outputs are slightly different:

```
## sf

qtm(countries_sf)
qtm(countries_sf, fill = "POP2005")

tm_shape(countries_sf) +
  tm_polygons("POP2005", palette = "Greens", style = "jenks")

## sp

qtm(countries_sp)
qtm(countries_sp, fill = "POP2005")

tm_shape(countries_sp) +
  tm_polygons("POP2005", palette = "Greens", style = "jenks")
```

The qtm() and tm_shape() functions handle the data differently, particularly the default styles chosen when creating maps.

There are very few differences between the code for mapping using tmap, which is very helpful for us. However, not all geospatial libraries can plot data in both formats, so be aware of this.

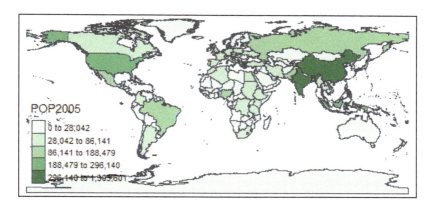

Figure 7.1: Output from tmap with sf data.

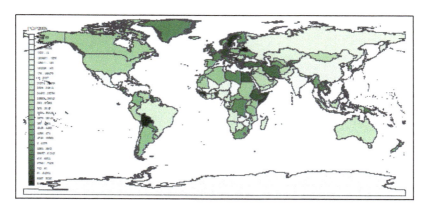

Figure 7.2: Output from tmap with sp data.

Plotting maps before tmap

Previously we had the raster package, but this is being replaced over time. Much like how sf is replacing sp for vector data, raster data is having a similar change. There are actually two new packages worth mentioning - terra and stars.

terra is more the direct replacement for the raster package. It is significantly faster than raster for many operations and has more long-term support. We will use terra in this practical and is the one I would recommend for new users. Most raster functions have equivalents in terra and it works is more or less the same way. There is a nice write-up of the differences between the two.[a]

One annoyance is that the terra package will not directly work with sf data types. There are convertor functions to convert from sf to SpatVector (which is the vector format terra uses) so it is not a major problem, it just makes the code a bit more clunky. Some code will work with sf data types, but it is a bit hit-and-miss, so you are probably going to need to do some conversion at some point. There is some interesting discussion on GitHub.[b]

The other package worth mentioning is stars. This performs a similar function to terra but is a bit more complex to use. There are some good guides available, and stars can deal with more complex data types that terra. One particular example is it handles a variety of grid formats, including curvilinear grids, which are very common in climate science, particularly NetCDF files. See GitHub for more details, including examples.[c]

---

[a]https://oceanhealthindex.org/news/raster_to_terra/
[b]https://github.com/rspatial/terra/issues/89
[c]https://r-spatial.github.io/stars/articles/stars4.html

Before the tmap library was available, we could still plot maps, but it was a much more complex process:

```
#select variable
        var <- as.numeric(countries_sp@data$POP2005)
#set colours & breaks
library(classInt)
breaks <- classIntervals(var, n = 6, style = "fisher")
library(RColorBrewer)
my_colours <- brewer.pal(6, "Greens")
#plot map
plot(countries_sp, col = my_colours[findInterval(var,
breaks$brks, all.inside = TRUE)], axes = FALSE,
border = rgb(0.8,0.8,0.8))
        #draw legend
library(maptools)
legend(x = -190.5203, y = 25.23282,
legend = leglabs(breaks$brks),
            fill = my_colours, bty = "n", cex = 0.65)
```

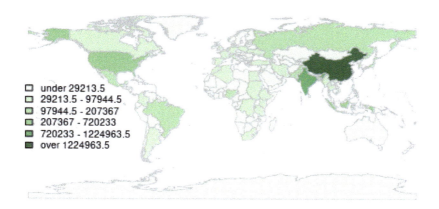

Figure 7.3: Output from base R of sp data.

7.4   More Advanced Analysis

The differences between sf and sp get bigger with the more advanced analysis. Now, all spatial analysis can be done in sf, so it's unlikely you will need sp.

7.5   Future Developments

sp and related packages (rgdal, rgeos, maptools) have been retired at the end of 2023 with them being replaced by the sf, terra and stars packages. There is a very good write up of the plan moving forward,[28] and there will be more posts on r-spatial.org over time. There is also a very good comparison of sp commands and the equivalent in sf on GitHub.[29]

Future plans will also be shared on the blog at r-spatial.org and email list.[30]

---

[28]https://r-spatial.org/r/2022/04/12/evolution.html

[29]https://github.com/r-spatial/sf/wiki/Migrating

[30]https://stat.ethz.ch/mailman/listinfo/r-sig-geo

Part 3

Advanced Concepts

8. Loops

8.1 Learning Outcomes

After reading this chapter, you will:

- Understand how we can use loops to create multiple maps.
- Be able to adapt the loop code to choose which maps you want to create.

So far we have created a range of maps in R, including using vector data to create a choropleth map in Chapter 4. We can develop this code to:

a) save the map using code rather than the GUI tools, and

b) embed this code within a loop to allow us to create multiple maps quickly and easily.

8.2 Saving the Map in Code

We can tweak the code to export the map to a file, rather than showing it in the Plots window. This code will do that by saving the map as a new variable (m) and then save it using tmap_save(m):

```
#create map
m <- tm_shape(LSOA) +
tm_polygons("AllUsualResidents", title = "All Usual Residents",
      palette = "Greens", style = "equal") +
tm_scale_bar(width = 0.22, position = c(0.05, 0.18)) +
tm_compass(position = c(0.3, 0.07)) +
tm_layout(frame = F, title = "Liverpool", title.size = 2,
          title.position = c(0.7, "top"))
#save map
tmap_save(m)
```

This will save the file as tmap01.png in your working directory:

```
> tmap_save(m)
Map saved to /home/nickbearman/GIS/using-r-gis/tmap01.png
Resolution: 3054.149 by 1443.938 pixels
Size: 10.1805 by 4.813125 inches (300 dpi)
```

> **Important:** It is important to remember that these commands will save the map as a PNG in your working directory. It will NOT show the map in the Plots window like we have been used to. So make sure you check your working directory to see the map.

We can also customise the filename by replacing that line of code with this:

```
tmap_save(m, filename = "map-Pop2005.png")
```

This is really useful because it allows us to use the code to save the map, rather than use the graphic interface. This opens up a wide range of possibilities, including creating multiple maps at once.

8.3   Creating Multiple Maps at Once

Now that we've written the code to create one map and save it, we can extend this easily to create multiple maps. This is one of the big advantages of using a code-based method to create maps, rather than a graphic interface. We can make use of computer programming concepts such as loops, which allow us to repeat a section of code multiple times, whilst changing thdata shown.

If we were using a graphic interface, we would need to manually create each map in turn. Not a problem if we had 4 or 5 maps, but it would quickly become quite trying if we had 10, or 100 maps to do!

> There are tools in QGIS and ArcMap that can replicate this process to create multiple maps. However, it is a more complex process to go from one map to multiple maps, whereas with a loop in R, it is easier.

So far we have our script to save a single map as a file.

```
m <- tm_shape(countries) +
  tm_
polygons("POP2005", title = "POP2005", palette = "Greens",
              style = "equal") +
  tm_scale_
bar(width = 0.22, position = c(0.05, 0.10)) +
  tm_compass(position = c(0.3, 0.07)) +
  tm_
layout(frame = F, title = "Population 2005", title.size = 2,
            title.position = c(0.7, "top"))

#save map with a specified filename
tmap_save(m, filename = "map-Pop2005.png")
```

Make sure you are happy with this before we continue. If you want to go back and check Chapter 4 for more details, please do so.

To make this into a loop, this is the code we need to change it into:

```
#set which variables will be mapped
mapvariables <- c("POP2005", "LIFEEXP", "LIFEEXPM")

#loop through for each map
for (i in 1:length(mapvariables))
#setup map

m <- tm_shape(countries) +
#set variable, colours and classes
tm_polygons(mapvariables[i], palette = "Greens",
  style = "equal") +
#set scale bar
tm_scale_bar(width = 0.22, position = c(0.5, 0.01)) +
#set compass
tm_compass(position = c(0.8, 0.05)) +
#set layout
tm_layout(frame = F, title = "Country Data", title.size = 2,
          title.position = c(0.65, "top"))
```

```
#save map
tmap_save(m, filename = paste0("map-",mapvariables[i],".png"))
#end loop
```

The only lines we've added or changed are shown in **bold**. The other lines haven't changed at all - they're exactly the same as in our previous example.

The first section is where we list which variables we want to map, i.e. which columns within the `countries` data frame we want to create maps for. Here we have listed 3: `"POP2005"`, `"LIFEEXP"`, `"LIFEEXPM"`. We could list 5, 50 or 500 here!

Next, we start the loop with `for (i in 1:length(mapvariables)) {`

A key bit of any loop is the `i` variable. This is a counter, or iterator, it counts how many times we have gone through the loop. It could have any name - `i` is just a convention. We could call it `apple` or `banana` if we wanted to.

The code then says go through the loop from 1 to `length(mapvariables)`, so however many elements are in `mapvariables`.

So for our example, it will be for `(i in 1:3) {`, go through the loop 3 times, starting at 1.

The then `{` starts the loop.

If you look at the end of the code you will see a `}` which is the end of the loop. All the code within `{` and `}` will be repeated.

This is the code that will be repeated:

```
#setup map
m <- tm_shape(countries) +
  #set variable, colours and classes
  tm_polygons(mapvariables[i], palette = "Greens", style = "equal")
  #set scale bar
  tm_scale_bar(width = 0.22, position = c(0.5, 0.01)) +
  #set compass
  tm_compass(position = c(0.8, 0.05)) +
```

```
#set layout
tm_layout(frame = F, title = "Country Data", title.size = 2,
          title.position = c(0.65, "top"))
#save map
tmap_save(m, filename = paste0("map-",mapvariables[i],".png"))
```

The first time through, i will be replaced with 1.

So mapvariables[i] will be replaced with mapvariables[1] which is POP2005. Then the code is:

```
#setup map
m <- tm_shape(countries) +
  #set variable, colours and classes
  tm_polygons(POP2005, palette = "Greens", style = "equal") +
  #set scale bar
  tm_scale_bar(width = 0.22, position = c(0.5, 0.01)) +
  #set compass
  tm_compass(position = c(0.8, 0.05)) +
  #set layout
  tm_layout(frame = F, title = "Country Data", title.size = 2,
            title.position = c(0.65, "top"))
#save map
tmap_save(m, filename = paste0("map-",POP2005,".png"))
```

which is very similar to our example above.

Then R gets to the end of the loop, and then repeats it, but with i = 2, which is mapvariables[2] which is "LIFEEXP", which gives us this code:

```
#setup map
m <- tm_shape(countries) +
  #set variable, colours and classes
  tm_polygons(LIFEEXP, palette = "Greens", style = "equal") +
  #set scale bar
  tm_scale_bar(width = 0.22, position = c(0.5, 0.01)) +
  #set compass
  tm_compass(position = c(0.8, 0.05)) +
```

```
#set layout
tm_layout(frame = F, title = "Country Data", title.size = 2,
          title.position = c(0.65, "top"))
#save map
tmap_save(m, filename = paste0("map-",LIFEEXP,".png"))
```

The only differences between the two chunks of code are in bold.

R will continue doing this as long as there are values in `mapvariables`. When it has done all 3, it will stop the loop and move on to the next bit of code.

Loops are really powerful tools in R and allow us to create multiple maps quickly, and it is easy to adapt existing code into a loop. They aren't the easiest concept to explain in a book, so do check out my video as well, from my training course, which might make some bits clearer.[31]

8.4   Exercise - Custom Map Titles

Currently, we have the same title on each map which is not particularly useful. See if you can adapt the code to generate a different title for each map. There are a few different ways of doing this. There is an 'answer' in the script file on the website, but try not to look before you've had a go at working it out yourself!

---

[31]https://youtu.be/l_op2SDenU4

9. If, If Else & Functions

9.1 Learning Outcomes

After reading this chapter, you will:

- Be aware of more advanced code flow functions like `if` and `else`.
- Know how to use these within R code.
- Be able to write a basic function.

One of the big advantages of using R as a GIS over a typical graphical interface GIS is the fact that we use scripting to create maps. This means that we can make use of programming concepts to run our code, and therefore to do our analysis. We've already seen how to use loops to make multiple maps in Chapter 8. However, loops aren't the only thing we can use for this, we can also use `if` and `else` statements, along with functions. These are very useful in different situations, and by the end of the chapter you will know how to use them yourselves.

9.2 If Else Statements

If Else statements allow us to control which bit of code runs depending on an option or variable we specify. If you've ever used If statements in Excel, the approach is much the same. If you haven't, don't worry - I'm confident you will pick up the idea quickly.

To give you an idea of how it works, I'm going to give you an example in what is sometimes called "pseudo code" - plain English mixed with some programming terms. We might have some code like this:

```
If weather = sunny then
        Take sun hat and sun cream
Else if weather = rainy then
        Take umbrella
Else if weather = cold then
        Take hat and gloves
```

```
Else then
        Take waterproof coat
```

What this does is look at the variable weather and makes a decision about what action to take based on what the value of that variable is. It's worth noting that in this example weather can only take one value. It is not possible for it to be sunny and cold at the same time in our example. This is a very common approach when modelling data - we have to simplify the real world for a computer to understand. Whether the model is too simple to actually be useful is an important thing to consider when making use of the results from a model.

Anyway, in our case, the code above is saying that if the weather is sunny, then take a sun hat and sun cream. The third line starting else if is giving another option, which allows us to give a second condition (if weather = rainy in this example). We might summarise that as saying if the weather is sunny, then take sun hat and sun cream, otherwise if the weather is rainy, then take the umbrella. Else if is the way we join these statements together, so only one of the lines starting with take is processed. In this example, we have two of these else if options, but you can include as many as you want to.

The final else line is slightly different. In this case, it means if weather has any other values (apart from the ones covered already, i.e sunny, rainy or cold), then take your waterproof coat. I've written this with the UK in mind, where if it is not raining at the moment, it is likely to be raining later!

9.3   If Else Statements in R

So this is all very well, but how do we implement this in R? The first step is to think about what we want to do. Sometimes people like to write this out in pseudocode to think about the 'logic' - i.e. the different steps the computer needs to go through. Here we're thinking about whether the spatial data we have is raster or vector, and then supply different code to map it:

```
Read data in
If data = vector
        Then use `tm_shape` function from the `tmap` library to map the da
```

```
Else if data = raster
        Use R base `plot()` function to map the data
```

So this is our initial outline of what we want the code to do. This process helps us think through the logic of the different steps. So, I had a go at writing the code, and came up with a problem! You may have spotted it already - the first line is Read data in. But what command we use depends on whether the data are vector or raster. For example, the command to read in vector data is:

```
library(sf)
data <- st_read("world_countries.shp")
```

However, the command to read in raster data is:

```
library(terra)
data <- rast("KEN_population_v1_0_mastergrid.tif")
```

So how do we know what data we have? To know what type of data we have, we need to read it in and look at it, but in order to read it in we need to know what type of data we have!

In the end, for this example, I cheated slightly and decided we could guess the data based on the filename. So, if our filename ends in .shp then we can guess it is a vector format, or if it ends in .tif then we can guess it is a raster format. This isn't ideal - because we can get vector data that is not a shapefile, and raster data that is not a tiff file. However, it will work for this example. This is also a good opportunity to add an else statement to do something different if the filename doesn't end .shp or .tif, such as show an error message. My updated pseudocode is:

```
Set data filename
If data filename ends in `.shp` then data is vector
        Use `st_read()` to read in the data
        Then use `tmap` library to map the data
Else if data filename ends in `.tif` then data is raster
        Use `rast()` to read in the data
        Use R base `plot()` function to map the data
Else, the data filename ends in something else
        Show an error message
```

> **Remembering Syntax**
>
> As you know, R has a particular way of writing code in terms of layout, how things are structured, whether to use " or ', brackets, etc. Different programming languages have different approaches for this, and for example, Python doesn't generally use brackets, but uses tabs instead. This is called the syntax of a language.
>
> When I am writing R code, I can remember how an If statement works, but I always have troubling remembering the syntax - i.e. exactly how it is laid out. This is very common, so I just ended up googling 'R If Else statement' and I found an example::
>
> ```
> if (test-expression1) {
> statement1
> } else if (test-expression2) {
> statement2
> } else if (test-expression3) {
> statement3
> } else {
> statement4
> }
> ```
>
> This helped me work out how to write my code. When you're writing code, don't expect yourself to know everything. Use your skills to look up what you need to know. This is common practice and completely normal.

Now we can write this out as R code:

```
#set data_filename
data_filename <- "world_countries.shp"
#data_filename <- "nga_ppp_2020_1km_Aggregated.tif"

#extract last 3 characters of filename to determine filetype
format <- substr(data_filename,nchar(data_filename)-2,nchar(data_filename)

if ( format == "shp") {
  #map vector data
  data <- st_read(data_filename)
  tm_shape(data) +
       tm_polygons("POP2005", palette = "Greens", style = "jenks")

} else if ( format == "tif") {
  #map raster data
```

```
library(terra)
data <- rast(data_filename)
plot(data)

} else {
print("I don't know how to map this data.")
print("It is not a shapefile (*.shp) or a tiff file (*.tif)")

}
```

There's a few things worth noting. Starting in the middle, we have this code for vector:

```
#map vector data
data <- st_read(data_filename)
tm_shape(data) +
  tm_polygons("POP2005", palette = "Greens", style = "jenks")
```

And this code for raster:

```
#map raster data
library(terra)
data <- rast(data_filename)
plot(data)
```

This is the same code as we used in Chapter 4 (vector) and Chapter 5 (raster) so all is fine so far. The only difference is that I've used data_ filename as a variable instead of the actual data filename. Exactly the same principle as we used in the loop code in Chapter 8. On line 1, we set data_filename to whatever we want it to be. In this example, line 1 sets it to some vector data, and line 2 sets it to some raster data. Currently, I've commented-out line 2 (using the #).

For the if statement itself, note the brackets. Normal brackets () for the condition statement - what we're checking to see whether we run that section. format is a variable set using the substr() function, based on the last three letters of the filename (see below).

Also note that we use == two equals signs. In R, this is the comparative operator, i.e. we're saying does a equal b? (or does format equal shp in this case).

In R, one equal sign means assign. It can be used instead of <-, for example:

```
data <- st_read(data_filename)
```

and

```
data = st_read(data_filename)
```

are identical. Both mean set data to be equal to st_read(data_filename). This is the syntax of R, using = to assign and using == for comparison. Some schools of thought happily use = for assign but you may have noticed that I don't. I always use <- for assign. This is because this was how I was taught, and to me, it seems clearer to use <- for assign and == for comparison. I've stuck with this, and avoided using = at all to minimise confusion.

Line 4 is also worth spending a couple of minutes on explaining - this is how we achieve the task of working out the file type, by looking at the last three characters of the filename. The substr() (substrings) function is very useful for this. To expand that line a bit:

```
format <-
        substr(
                data_filename,
                nchar(data_filename)-2,
                nchar(data_filename)
                )
```

The function substr() has three variables. The first one is data_filename which is the text string it is extracting values from. The last two are the character positions of the first and last characters we want to extract. For example, if we wanted to extract the text from the 3rd to the 5th position from hello the command would be:

```
substr("hello", 3, 5)
```

And the output would be:

```
llo
```

In our case, we don't know how long the filename will be (because it can vary) but we know we want the last three characters. So we use the nchar() (number of characters) function to calculate the number of characters in data_filename. The start number is two less than the length (remember in our example 5-3 = 2), and the end number is the length of data_filename.

Finally, we have an error message that R will print if the last three characters of the filename aren't shp or tif. This is something called error handling and is often useful in these types of situations.

9.4   Writing Your Own Functions

We can also take this code a step further and create a custom function. For example, we could take our mapping code and turn in to a function that would be easy to use. This code is identical to the if else code above, apart from the first and last lines:

```
Mapping <- function(data_filename) {
        #extract last3 characters of filename to determine filetype
        format <- substr(data_filename,nchar(data_filename)-2,
                nchar(data_filename))

        if ( format == "shp") {
                #map vector data
                data <- st_read(data_filename)
                tm_shape(data) +
                tm_polygons("POP2005", palette = "Greens", style = "jenks")

        } else if ( format == "tif") {
                #map raster data
                library(terra)
                data <- rast(data_filename)
                plot(data)

        } else {
                print("I don't know how to map this data.")
                print("It is not a shapefile (*.shp) or a tiff file (*.tif)")
        }
}
```

The first line gives the name of the function Mapping and the parameters that it uses. Our function is quite simple and only has one parameter - data_filename. It also has an open curly bracket { which all the function code goes in. The final line is } which indicates the end of the function.

If we run our function in R, R stores it as a custom function we can use anytime. You will see it listed in the environment under Functions. We can then run it like this:

```
Mapping("world_countries.shp")
```

This will take our input variable ("world_countries.shp") and create the map. Try it with the raster filename too.

We can get more advanced with functions, for example, to do specific analyses or other processes. It can be really useful when you have a set of code that you have to repeat frequently with different data, or with different variables. We could do a similar approach with the loop example earlier:

```
MappingColumn <- function(data, column_name) {
#setup map
m <- tm_shape(countries) +
        #set variable, colours and classes
        tm_polygons(column_name, palette = "Greens", style = "equal") +
        #set scale bar
        tm_scale_bar(width = 0.22, position = c(0.5, 0.01)) +
        #set compass
        tm_compass(position = c(0.8, 0.05)) +
        #set layout
        tm_layout(frame = F, title = "Country Data", title.size = 2,
                    title.position = c(0.65, "top"))
#save map
tmap_save(m, filename = paste0("map-",column_name,".png"))
#end function
}
```

Here we have replaced mapvariable[i] with column_name and countries with data. We can then run it with whichever column name we like:

```
MappingColumn(countries, "POP2005")
MappingColumn(countries, "LIFEEXP")
```

Of course, the column name needs to exist, and if it doesn't, we will get an error.

We could add an if statement to check whether the column exists. This is an area of function writing called error checking and is an important element if you want to share your function with other people.

One other important aspect worth mentioning is that libraries in R are merely a collection of different functions. So if you have a function (or series of functions) that you want to share with others, you can put them together as a library. Initially, you can share it on GitHub (you may have installed functions from GitHub before) and ultimately you can ask for your library to be included in CRAN, the central list of all R libraries. More details on this are beyond the scope of this book, but if you want to share code, it is easy to do!

10. Spatial Data Wrangling

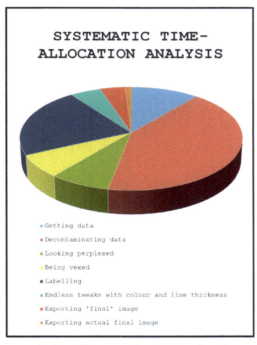

SYSTEMATIC TIME-
ALLOCATION ANALYSIS

- Getting data
- Decontaminating data
- Looking perplexed
- Being vexed
- Labelling
- Endless tweaks with colour and line thickness
- Exporting 'final' image
- Exporting actual final image

Figure 10.1: How much time different tasks in GIS takes, from undertheraedar on Twitter.

10.1   Learning Outcomes

After reading this chapter, you will:

- Be aware of a number of techniques to get your data in usable format.
- Know how to search effectively for help.
- Know how to ask effectively for help.

145

Once you have your data, you may need to do some data wrangling to get it in a suitable format. This could be removing outlying points, dealing with incorrectly located points, removing unnecessary columns or generalising or clipping data, so it is easier to process. Whilst this 'data decontamination' (as it is sometimes termed) can be quite time-consuming, it is worth while - because it will allow you to complete your analysis and be confident of the output you've created. Sometimes in the world of data science, this is called 'data wrangling'.

There is no 'one size fits all' approach to this - so much depends on what data you have and what you're trying to do with it. This chapter will cover a few useful techniques, but you will need to decide if they're relevant & useful for your own data.

10.2   Reading your Non-spatial Data in

First of all, you need to read your data in. If you're working with CSV / TXT data, the read.csv() function is your friend. It has a few options, but generally works well.

We're going to use some of the population data we were working with in Chapter 4, and some additional data. Download ch10-data.zip from the book website and read in pop2015.csv

```
data <- read.csv("pop2015.csv")
```

If you are working in a language that uses , to separate decimal points rather than . then you may need to use read.csv2():

```
How you show your number      | which Library
1,234,567.89                  | read.csv()
1.234.567,89                  | read.csv2()
```

Techy bit: CSV stands for 'comma separated values' and if you look at the source of a CSV file (e.g. open it in Notepad) your data that looks like this:

```
Name   | Number | Decimal Number
Cait   | 12     | 56.78
James  | 34     | 34.78
Mike   | 45     | 38.9
```

will look like this:

```
Name,Number,Decimal Number
Cait,12,56.78
James,34,34.78
Mike,45,38.9
```

If you use a comma to separate numbers (e.g. 56,78) then you can see where the problems start to come from. R will be expecting 3 columns on line 1, but 4 columns on lines 2-4 in the example above. read.csv2() solves this by explicitly saying to expect , in numbers.

10.3 Reading your Spatial Data in

If you are working with spatial data (shapefiles, geopackages, etc.) then you will probably be using the st_read() function. Usually this works quite well, but sometimes it has issues.

Here are some common problems:

10.3.1 could not find function "st_read"

```
> countries <- st_read("world_countries.shp")
Error in st_read("world_countries.shp") :
```

```
could not find function "st_read"
```

This error means that R can't find the function `st_read()`. Odds are you need to re-load the library (`sf`). Run `library(sf)` and try again.

10.3.2   Error: The file doesn't seem to exist

```
> countries <- st_read("world.shp")
Error: Cannot open "world_countries_update.shp";
  The file doesn't seem to exist.
```

If you get this message, it means that R can't find the file. There are two things to check - is the file name right, and is the working directory correct?

You can check the working directory in RStudio in two ways. You can see on screen what the working directory is set to (see figure 10.2):

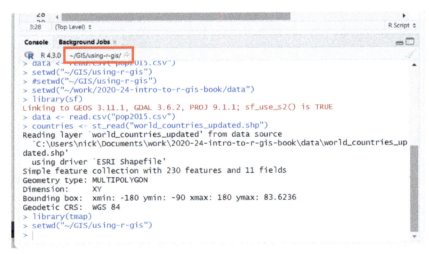

Figure 10.2: The current working directory, highlighted in red, shown just above the Console window.

You can also use the command `getwd()` to get the path to the working directory.

```
> getwd()
[1] "C:/Users/nick/Documents/GIS/using-r-gis"
```

Remember that when opening a file, the file must be in the working directory itself - not in a sub-directory. So for example, if the file was in a data folder within using-r-gis it would not work. The file would need to be in the folder using-r-gis.

10.3.3   Shapefile not complete

```
> countries <- st_read("world_countries_updated - Copy.shp")
Error: Cannot open "C:\Users\nick\Documents\work\
2020-24-intro-to-r-gis-book\data\world_countries_updated - Copy.shp";
The source could be corrupt or not supported. See `st_drivers()` for a
list of supported formats.
In addition: Warning message:
In CPL_read_ogr(dsn, layer, query, as.character(options), quiet,  :
  GDAL Error 4: Unable to open C:\Users\nick\Documents\work\
     2020-24-intro-to-r-gis-book\data\world_countries_updated - Copy.shx
     or C:\Users\nick\Documents\work\2020-24-intro-to-r-gis-book\data
     \world_countries_updated - Copy.SHX. Set SHAPE_RESTORE_SHX config
     option to YES to restore or create it.
>
```

This is an interesting one - the first section means it cannot read the file, but on line 4 it says:

```
Unable to open C:\Users\nick\Documents\work\2020-24-intro-to-r-gis-book\
data\world_countries_updated - Copy.shx
```

This means it can't find the .shx part of the shapefile. Remember from Chapter 2 that a shapefile has at least three files. In fact the .shx file is an index file - it is theoretically possible to recreate this, but it means delving into the GDAL library to do so. There is no way to do this from within R (as far as I can find - if you know differently, let me know!).

In this case, have a look where you downloaded the file from and make sure you have all the component parts of the file - *.shp, *.shx and *.dbf are the absolute minimum.

These are the most common issues with reading files. If you find any more, please do post questions on the book website!

10.4   Viewing Head/Tail

Once you've read your data in, the first thing to check is whether it is what you were expecting. There are a number of things you can do - firstly, how many rows and columns does it have? You can use the nrows() and ncols() functions for this, or see it in the RStudio Workspace (see figure 10.3):

```
> nrow(data)
[1] 228
> ncol(data)
[1] 3
```

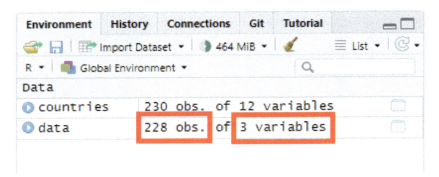

Figure 10.3: The Environment tab in RStudio, showing the data variable with 228 observations (rows) and 3 variables (columns).

Secondly, you can either click on the data listed in the environment, or run the View() command to open the data in a new tab and have a look. This works well for small data sets (<~100 rows, <~10 columns) but can be slow with larger data sets. If you run the command and the data takes more than about 10 seconds to show, or the tab appears, but it is blank, the data is probably too large to view this way.

With larger data sets you can use the head() or tail() functions to see the first (or last) 6 rows of data. You can combine this if you like the View() functionality:

```
View(head(large_data))
```

Here you can check if the column contents and column names make sense and match. If it is a small data set, you can also sort the data - click on the column name. Sometimes this takes a few seconds, even with quite small data like our countries data. Be patient!

If you're working with large datasets, then sometimes we have to work with R in a particular way. There are a few approaches, and which is best depends on your particular data and circumstances.

- This is a great overview of working with large data in R in general[32]
- When it comes to spatial data, there are a couple of additional points. One great approach is to store our spatial data in a database. This isn't something we will cover in detail here, but see r-spatial.org for some good suggestions.[33]

Or sometimes R just won't work with very large data sets and you will need to cut down your data in another program. Remember this as an option - often you may not need all the entries in your data, or all of the columns, and cutting them down can help a lot.

If you're working with spatial data, you can do all the above (head(), tail() etc.) on the attribute table, but it is also worth checking the spatial element when you read data in. The qtm() function works well for this for vector data:

```
qtm(countries)
```

While R has many advantages, quickly viewing spatial data is not one of them. Sometimes I find it really useful to load up spatial data in a graphical GIS, e.g. QGIS, to have a quick look - so if this works for you too, then give it a go.

---

[32]https://hbs-rcs.github.io/large_data_in_R
[33]https://r-spatial.org/book/09-Large.html

10.5   Incorrect Geometries

Sometimes your shapefile will read in correctly, but when you try to
map it you might get this error message:

```
> countries <- st_read("world_countries.shp")
Reading layer 'world_countries' from data source
  'C:\Users\nick\Documents\work\2020-24-intro-to-r-gis-book\data\
    world_countries.shp'
  using driver 'ESRI Shapefile'
Simple feature collection with 230 features and 11 fields
Geometry type: MULTIPOLYGON
Dimension:    XY
Bounding box: xmin: -180 ymin: -90 xmax: 180 ymax: 83.6236
Geodetic CRS: WGS 84
> library(tmap)
> qtm(countries)
Error: Shape contains invalid polygons. Please fix it or set
tmap_options(check.and.fix = TRUE) and rerun the plot
```

Error: Shape contains invalid polygons. Please fix it or set
tmap_options(check.and.fix = TRUE) and rerun the plot

This error means there is something wrong with the geometry of the
polygons. The geometry is how the polygons are stored in our data.
Usually, polygons are stored as a series of points in a specified order to
enclose an area. This error invalid polygons means there is something
wrong with how the polygons are stored.

R is very picky about how polygons are stored. If you tried opening
this file in QGIS, it would probably work ok - give it a try and see what
happens. There are a few ways of fixing this invalid geometry. First of
all you can try tmap_options(check.and.fix = TRUE) as the error sug-
gests. In our case, this doesn't help:

```
tmap_options(check.and.fix = TRUE)
qtm(countries)
Warning message:
The shape countries is invalid. See sf::st_is_valid
```

What's the ':::' about?

I wrote out the st_is_valid() function as sf::st_is_valid() so what does the :: mean? In R, this is a way of specifying the library as well as the function. Some function names are quite generic, so if you are using multiple libraries, it can be difficult to know which library you are referring to. With the :: method you specify the library and then the function name as a way of explicitly saying which library and function you are talking about.

If we look at the helpfile for sf::st_is_valid:

```
?sf::st_is_valid
```

We have two options - st_is_valid and st_make_valid.

The st_is_valid function will tell us whether the geometry is valid. The st_make_valid function can fix some of these problems:

```
st_is_valid(countries)
```

Calling st_is_valid(countries) will give us a list of TRUE or FALSE for each polygon, and whether it is valid or not:

```
st_is_valid(countries, reason = TRUE)
```

will give us a slightly more useful output telling us what is wrong. Adding table() will summarise the results by the reason for the incorrect[34] geometry:

```
table(st_is_valid(countries, reason = TRUE))
```

Another approach we can take to making the geometries valid is running:

---

[34]If you are interested in the gory details of what makes geometry incorrect, I recommend this blog post: https://r-spatial.org/r/2017/03/19/invalid.html

```
countries2 <- st_make_valid(countries)
```

Then running:

```
table(st_is_valid(countries2, reason = TRUE))
```

Shows us they are all valid now.

We can then plot the map as usual:

```
qtm(countries2)
```

If you can't fix the geometries using the methods above, there are a couple of other things you can try.

- Firstly, you can see if the file will open in QGIS (or another GIS program). QGIS is a bit more relaxed about the requirements for geometry compared to R, so will tolerate some minor errors. If it does work, then you can export the file from QGIS and try loading that in R.
- Secondly, you can try creating a very small buffer (e.g. 0.01 meters) around the polygon. You can go this in R (using the st_ buffer function) or QGIS.
- The final option is to see if you can find a different copy of the spatial data from wherever you got it from. It might be available at a different generalisation level, or in a different format.

10.6 Converting Data Types

Sometimes when reading data in to R from a CSV file, R will have to make the decision about whether the data is a number or is text. It usually gets this right, but occasionally it doesn't. This is particularly important when joining data, as we did back in Chapter 4. R is usually pretty good at storing data as the right type, but you can get into situations where you read in a CSV file, and the numerical data is stored as strings or factors. This is when we need to intervene.

> **Factors**
>
> Factors are a slightly odd data type in R, not seen often in other software. They're designed to be used in multiple choice surveys, and a factor is one value out of a series of specified possible values. This is important to know for some types of survey analysis, but not much help for us. This used to be a very common issue, as R tended to read in data as factors by default. There was even a parameter (stringsAsFactors) which had to be set as FALSE to solve this. From version 4.0.0 or R released in April 2020, stringsAsFactors will be set to TRUE as default, so this is no longer an issue. See https://blog.r-project.org/2020/02/16/stringsasfactors/ for a nice write up.

There are two key data types to be aware of here - numbers (numeric) and text (string). If you try and join data in different data types, R will give you an error message:

```
> countries$NAME[1] + countries$NAME[2]
Error in countries$NAME[1] + countries$NAME[2] :
  non-numeric argument to binary operator
```

This is the thing to look out for, and means we need to check what data type each column is. We can use either the class() function to check the type of a specific column, or the str() function to check a whole data frame.

```
        str(countries)
```

```
> str(countries)
Classes 'sf' and 'data.frame':       230 obs. of  12 variables:
$ FIPS    : chr  "AC" "AG" "AJ" "AL" ...
$ ISO2    : chr  "AG" "DZ" "AZ" "AL" ...
$ ISO3    : chr  "ATG" "DZA" "AZE" "ALB" ...
$ UN      : int  28 12 31 8 51 24 16 32 36 48 ...
$ NAME    : chr  "Antigua and Barbuda" "Algeria" "Azerbaijan" ...
$ AREA    : int  44 238174 8260 2740 2820 124670 20 273669 768230 ...
$ POP2005 : num  83 33268 8563 3082 3015 ...
$ LIFEEXP : num  75 73.1 70.1 76.3 74 ...
$ LIFEEXPM: num  72.6 71 67.1 73.4 70.6 ...
$ LIFEEXPF: num  77.4 75.2 73.1 79.7 77.3 ...
$ INFMRT  : num  10 34.2 41.1 16.1 21 ...
$ geometry:sfc_MULTIPOLYGON of length 230; first list element: List...
```

```
..$ :List of 1
.. ..$ : num [1:23, 1:2] -61.7 -61.7 -61.8 -61.9 -61.9 ...
..$ :List of 1
.. ..$ : num [1:25, 1:2] -61.7 -61.7 -61.7 -61.7 -61.8 ...
..- attr(*, "class")= chr [1:3] "XY" "MULTIPOLYGON" "sfg"
- attr(*, "sf_column")= chr "geometry"
- attr(*, "agr")= Factor w/ 3 levels "constant","aggregate",...
..- attr(*, "names")= chr [1:11] "FIPS" "ISO2" "ISO3" "UN" ...
```

Here R lists all of the columns in countries and tells us what data type they are:

- chr character (similar to a string)
- num numeric (any type of number)
- int integer (whole numbers only)

Here, they are as we expect - country names are characters, and the POP2005 data is numeric. int is integer and is used to represent whole numbers. num is more generic and can be used for whole numbers and decimal numbers.

If we compare this with the non-spatial data we are joining to:

```
str(data)
```

```
'data.frame': 228 obs. of  3 variables:
$ UN_Code: int  28 12 31 8 51 24 16 32 36 48 ...
$ Name   : chr  "Antigua and Barbuda" "Algeria" "Azerbaijan" "Albania" ...
$ POP2015: int  92 39667 9754 2897 3018 25022 56 43417 23969 1377 ...
```

Here we can see Name is the chr character type as well. So we will be able to join countries$NAME to data$Name as we did in the example in Chapter 4.

Here we have created an example where R reads everything as characters:

```
data <- read.csv("pop2015-new.csv", colClasses = "character")
```

Run str(data) to check.

We can still join the data:

```
countries <- merge(countries, data, by.x="NAME", by.y="Name")
```

But when we draw a map:

```
qtm(countries, fill = "POP2015")
```

```
Warning message:
Number of levels of the variable "POP2015" is 219,
which is larger than max.categories (which is 30),
so levels are combined. Set tmap_options(max.categories = 219)
in the layer function to show all levels.
```

With the qtm() function, R will assume it is a categorical data set - note how all the values are different colours, and the error message. It is saying that "POP2015" has more levels (different values) than the maximum number of categories (30), so it has combined some of these.

We can also get this issue when trying to create a choropleth map. If the data is stored as a string, R can't work with them. We can use the as.numeric function to convert the string to numeric, and save this as a new column:

```
countries$POP2015num <- as.numeric(countries$POP2015)
```

Then qtm will work as we expect:

```
qtm(countries, fill = "POP2015num")
```

One of the important things to remember is that R is forcing the variable to be a number. If there are some actual letters in that column, you will get a message, and R sometimes inserts NAN (Not A Number) instead.

When you use the to.numeric function, it is good practice to check that the data has been converted how you expected.

10.7 Subsetting

Regularly when you download a set of data, you only actually need a small proportion of it. Subsetting is the term for cutting down your data to what you actually need. It's helpful to do this as early on as possible, as it will help you keep the file sizes down, which is important for some spatial analysis, as the larger the files, the longer the analysis will take. You can subset by attribute - or by location.

Subsetting by attribute is the easiest, and if you're working with spatial data, you can use this if you have the right attributes.

With the sf *library this got a lot easier, compared with* sp. *See Chapter 7 for details.*

We use a combination of square brackets [ ] and comparative operators == to do the subsetting.

For example, with our countries data we can select out the UK:

```
UK <- countries[which(countries$NAME == "United Kingdom"),]
head(UK)
qtm(UK)
```

This code takes the variable *countries,* and selects the rows (using the [,] operator and before the comma means rows) which the NAME column equals United Kingdom. This is then saved to the new variable named UK. We then look at the variable using the head() and qtm() functions to check the data and draw a quick map.

As with the joining, we can also use codes rather than names, which minimises the risk of spelling mistakes:

```
UK_FIPS <- countries[which(countries$FIPS == "UK"),]
```

We can also select multiple countries. Initially we can use the BOOLEAN operator OR, signified in R as |. This is sometimes called a vertical bar or pipe, and is usually on the same key on your keyboard as the \. Hold down shift to access it:

```
multiple_countries <- countries[which(countries$FIPS == "UK" |
countries$FIPS == "FR" | countries$FIPS == "SP"),]
qtm(multiple_countries)
```

Note, there is a slightly odd logic to using OR rather than AND. We might think we want to say select "UK" AND "FR" AND "SP". But in fact, what we are saying is 'select the rows where NAME equals "UK" and where NAME equals "FR" and where NAME equals "SP"'. However the country name (for a specific row) will only be one of these - it can't be all three at once. So we have to use OR which allows us to make this multiple choice.

Note the odd geography - Spain includes the Canary Islands (those on the bottom left) and while it has one row in the table, is actually made up of more than one polygon. This is called a multi-part polygon.

Multi-Part Polygons

Multi-Part Polygons are special types of polygons that represent multiple objects but are stored as one row in our attribute table. A great example is Spain, which consists of the Spanish mainland, but also some islands in the Mediterranean and the Canary Islands.

We can split this out with this command:

```
Spain_Islands <- st_cast(Spain,"POLYGON")
```

Formally, we are changing the data type from 'MULTIPOLYGON' (which can store these multiple polygons in one row) to 'POLYGON' (which can't). This splits Spain out into 16 separate bits. You can then adjust them as you wish. It's worth mentioning that the attribute data is just copied across all the rows automatically. This may or may not be helpful, which is what the warning message is saying:

```
Warning message:
In st_cast.sf(Spain, "POLYGON") :
      repeating attributes for all sub-geometries
  for which they may not be constant
```

In our case, the Life Expectancy data might still be relevant (as an average) but clearly the area column is rubbish - so we should ignore (or delete) this. See this Stack Exchange question for more details: https://gis.stackexchange.com/questions/305734/splitting-multipart-polygons-to-single-part-in-r/.

Finally, we can also select out ranges of data that meet certain criteria. For example, we could select all countries that have a life expectancy of more than 80 years:

```
life_exp <- countries[which(countries$LIFEEXP > 75),]
```

We can also vary this with other operators:

- < less than
- > greater than
- <= less than or equal to
- >= greater than or equal to
- != not equal to
- & AND
- | OR

We can also combine multiple ones together:

```
multiple_countries <- countries[which(countries$LIFEEXP > 75 &
    countries$INFMRT > 10),]
```

So far, we've been using criteria to select out different countries. However, this is not always necessarily the best way. Sometimes it might be easier to visually select the polygons you want to keep (or delete). This is not something R can do very easily, but it is something we can do really easily in QGIS. Often this is the case for the more visual or interactive element of GIS, that it is easier to do in a desktop GIS with a graphic user interface (GUI).

QGIS works really well when you want to do a custom selection, i.e. you click on a specific number of countries that you want to remove. This works very well when you are dealing with a small number, e.g. <20.

10.8 dplyr / tidyverse

It is also worth mentioning the dplyr library (part of the tidyverse). This is a very common library used in data science for data wrangling, usually for non-spatial data. However, you can use the same techniques on sf spatial data, which can be really useful for some data-wrangling tasks. There are many tutorials available on dplyr - I would recommend the RStudio ones as a starting point.[35]

---

[35]A nice tutorial: https://cran.r-project.org/web/packages/dplyr/vignettes/dplyr.html and some more general information https://dplyr.tidyverse.org/.

10.9   Raster Data

We've already seen how we can crop raster data in Chapter 5. There are also a number of additional techniques that we can use to get the raster data into the format that we need. Some of the same principles apply as we've already discussed with vector data, but raster data is a little different as it doesn't have an attribute table. Raster data is a grid of cells, each with a series of values. We can perform different operations on these values.

10.9.1   Raster algebra

The first technique is called raster algebra. This is when we apply functions to the raster as a whole. For example, we could add two raster layers together (figure 10.4):

Figure 10.4: Raster algebra in action. These two rasters are added together. Note how each pixel is added to the one in the same location on the other raster layer, e.g. for the top left pixel $1 + 1 = 2$.

It is crucial that the rasters are the same resolution and size for this to work. What happens is each cell is added to the cell in the same location on the other raster.

In our example in R, we're working with some population data from the 2011 Census for Great Britain [8]. We can load the data, and then crop out the area of St. Helens.

```
library(terra)
age0_14 <- rast("5a_ascii_grid2011_Age_Age0_14.asc")
plot(sthelens_0_14)
```

So far, this is what we have already done in Chapter 5

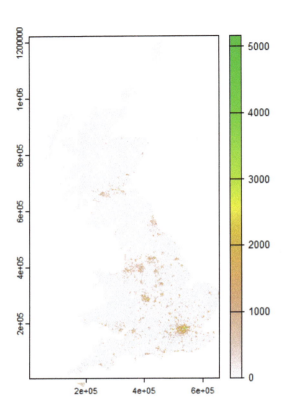

Figure 10.5: Plot of raster data sthelens_0_14.

```
# crop out a local authority
library(sf)
library(tmap)

sthelens <- st_read("sthelens.shp")

qtm(sthelens)

#convert to SpatVector
sthelens_spatVect <- vect(sthelens)

sthelens_0_14 <- crop(age0_14, sthelens_spatVect, mask = TRUE)
```

```
plot(sthelens_0_14)
plot(sthelens_spatVect, add = TRUE)
```

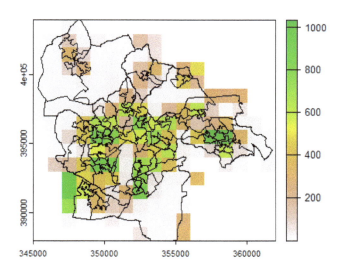

Figure 10.6: Plot of raster data sthelens_0_14 for St Helens.

In this section, we read in the St. Helens boundary, and then use this
to crop out the relevant section of the population data. With the terra
library, we have to convert the sf (shapefile) format data (sthelens) to
a format called SpatVector - this is a format used by the terra library.
Then we can use this to cut down the data.

In the crop function, we also use a mask - more on this later. Finally,
we plot the sthelens data plot(sthelens_0_14) and then also add the
vector boundary on top plot(sthelens_spatVect, add = TRUE). add =

TRUE means add this layer on top of the existing plot. By default, the plot() function will clear the existing plot first. The plot() function is a part of base R - not in a separate library. It is easy to use, but doesn't have the same number of features as tmap.

You might have also seen that R gave us an error message: Warning message: [crop] CRS do not match. This means that the CRS (coordinate reference systems) of the two data sets don't match. We can check them if we like:

```
> st_crs(age0_14)
Coordinate Reference System: NA
> st_crs(sthelens)
Coordinate Reference System:
  User input: Transverse_Mercator
  wkt:
PROJCRS["Transverse_Mercator",
    BASEGEOGCRS["GCS_OSGB 1936",
        DATUM["Ordnance Survey of Great Britain 1936",
            ELLIPSOID["Airy 1830",6377563.396,299.3249646,
                LENGTHUNIT["metre",1]],
            ID["EPSG",6277]],
        PRIMEM["Greenwich",0,
            ANGLEUNIT["Degree",0.0174532925199433]]],
    CONVERSION["unnamed",
        METHOD["Transverse Mercator",
            ID["EPSG",9807]],
        PARAMETER["Latitude of natural origin",49,
            ANGLEUNIT["Degree",0.0174532925199433],
            ID["EPSG",8801]],
        PARAMETER["Longitude of natural origin",-2,
            ANGLEUNIT["Degree",0.0174532925199433],
            ID["EPSG",8802]],
        PARAMETER["Scale factor at natural origin",0.9996012717,
            SCALEUNIT["unity",1],
            ID["EPSG",8805]],
        PARAMETER["False easting",400000,
            LENGTHUNIT["metre",1],
            ID["EPSG",8806]],
        PARAMETER["False northing",-100000,
            LENGTHUNIT["metre",1],
            ID["EPSG",8807]]],
    CS[Cartesian,2],
        AXIS["(E)",east,
```

```
        ORDER[1],
        LENGTHUNIT["metre",1,
            ID["EPSG",9001]]],
    AXIS["(N)",north,
        ORDER[2],
        LENGTHUNIT["metre",1,
            ID["EPSG",9001]]]]
```

Unfortunately, this is not the most helpful of outputs. The sthelens variable CRS is, in fact in British National Grid EPSG:27700. If you look carefully, you can probably spot GCS_OSGB 1936 near the top of the information. This is the datum that BNG uses, and if you see this, it is an indication that the data is probably in BNG. The output here is the conversion information for converting other systems to British National Grid. If you open the sthelens.shp file in QGIS, right-click on the layer, choose **Properties** and then **Information**, it will say (figure 10.7):

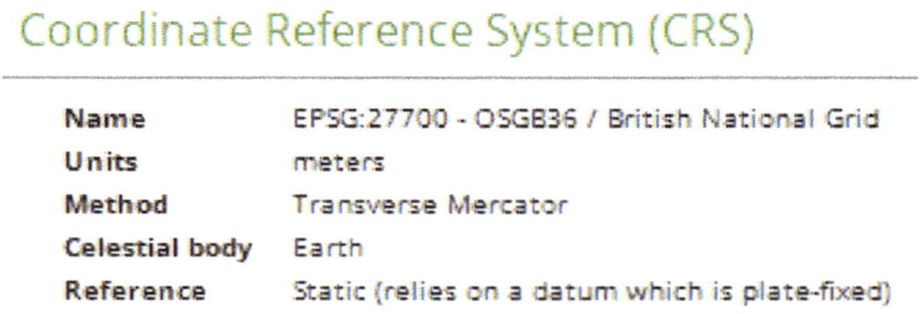

Coordinate Reference System (CRS)

Name	EPSG:27700 - OSGB36 / British National Grid
Units	meters
Method	Transverse Mercator
Celestial body	Earth
Reference	Static (relies on a datum which is plate-fixed)

Figure 10.7: QGIS Information on St Helens CRS.

Also, age0_14 doesn't have any CRS information attached to it, because the ASC (ASCII grid format) doesn't support this. However, it is actually also British National Grid. How do I know this? If we look at the website[36] and expand Coverage and Methodology and look under Spatial unit: it says Grid > British National Grid (see figure 10.8, on the next page).

There is most certainly an art to finding this information! This is something you will learn over time, the more you do this.

However, there is an easy way to confirm this:

---

[36]https://reshare.ukdataservice.ac.uk/852498/

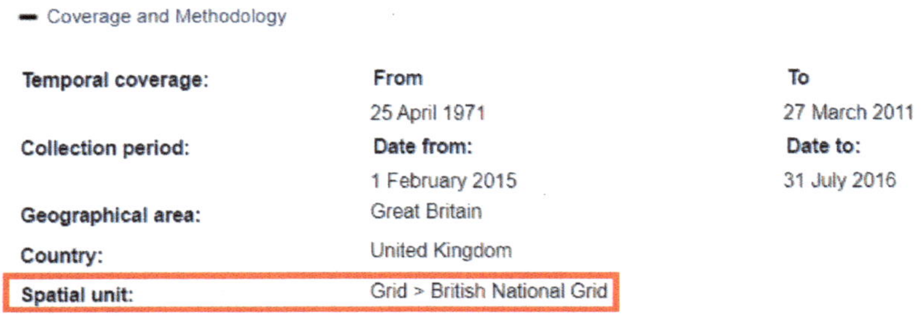

━ Coverage and Methodology

Temporal coverage:	**From**	**To**
	25 April 1971	27 March 2011
Collection period:	**Date from:**	**Date to:**
	1 February 2015	31 July 2016
Geographical area:	Great Britain	
Country:	United Kingdom	
Spatial unit:	Grid > British National Grid	

Figure 10.8: UK Data Service information on population data CRS.

```
plot(sthelens_0_14)
plot(sthelens_spatVect, add = TRUE)
```

When we plot the data, we can see it lines up correctly. Firstly, the data are in the same place, as if one data sets was in BNG and the other in Lat Long, there would be no overlap at all! Secondly, the coordinates are in the right format - 6 digits each. Finally, these units are LSOAs, each of which has about 1500 people in them. You can see the smaller ones have more people in them in each grid cell (green in colour) and the larger ones have fewer people in each grid cell (coloured red or white). So we can be confident both data are in British National Grid.

We can then repeat the process with numbers of people ages over 65:

```
age_65plus <- rast("5a_ascii_grid2011_Age_Age65p.asc")

age_65plus
plot(age_65plus)

sthelens_65plus <- crop(age_65plus, sthelens_spatVect, mask = TRUE)

plot(sthelens_65plus)

#note this now matches the sthelens area:

plot(sthelens_spatVect, add = TRUE)

#repeat for 65 plus
sthelens_65_plus <- crop(age_65plus, sthelens_spatVect, mask = TRUE)

#note the very different spatial pattern
```

```
plot(sthelens_65plus)
plot(sthelens_0_14)
```

It is not that easy to quickly compare data in RStudio. We can plot one dataset after another, but the later one will always replace it. We do have the arrows in the Plot window (figure 10.7, on page 165):

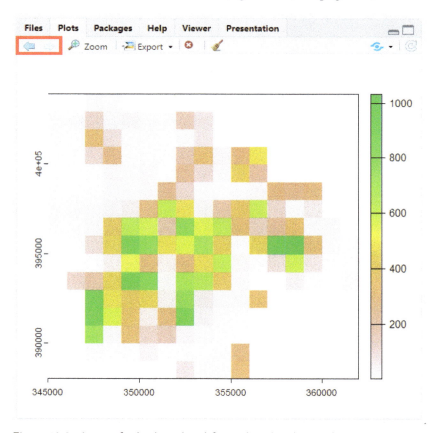

Figure 10.9: Arrows for backward and forward in the plot window.

Run the code above to plot the two maps. Then you can use the 'Back' arrow (pointing to the left) to go back to the previous plot (65 plus in our case). Then you can use the 'Forward' arrow to move to the next plot (0 to 14 in our case). This is a bit easier than re-running the code each time, and can allow a quick comparison.

We can now perform Raster Algebra on raster data. This involves adding two raster datasets together as mentioned earlier:

```
rasterA + rasterB = rasterC
```

For our data, note how the spatial pattern of the 65 plus and 0 to 14 data are very different:

```
sthelens_14_65 <- sthelens_0_14 + sthelens_65_plus
plot(sthelens_14_65)
```

Our output looks quite different to our input.

As well as addition, we can also apply other operators to raster data - subtract, multiply, divide and so on.

We can also divide one by the other:

```
sthelens_14_65 <- sthelens_65_plus / sthelens_0_14

plot(sthelens_14_65)
```

This is very commonly used to create indices by working with data with different wavelengths.

For example, we can take the red wavelength (R) and near infrared wavelength (NIR) channels from satellite data, and combine them like this to create the NDVI - Normalised Vegetation Differentiation Index [7].

```
(NIR - R) / (NIR + R)
```

If you're interested, here is a good working example using Landsat data: usgs.gov/landsat-missions/landsat-normalized-difference-vegetation index.

10.10   Reclassify Data

We also have the ability to reclassify certain values in a raster dataset. Working on our age data sthelens_65_plus, we have values from 0 to 992.

We can tell this by running:

```
sthelens_65_plus
```

Which gives us the summary data:

```
class       : SpatRaster
dimensions  : 16, 17, 1   (nrow, ncol, nlyr)
resolution  : 1000, 1000   (x, y)
extent      : 345000, 362000, 388000, 404000   (xmin, xmax, ymin, ymax)
coord. ref. :
source(s)   : memory
name        : 5a_ascii_grid2011_Age_Age65p
min value   :                    0.05251729
max value   :                  992.72650146
```

It's also worth looking at the histogram (figure 10.10, on the next page):

```
hist(sthelens_65_plus)
```

We can use the reclassify tool to group this data into more meaningful groups:

```
# let's reclassify:
# 0 - 100 low
# 100 - 500 medium
# 500+ high

m <- c(0, 6, 1,
       6, 13, 2,
       13, 25, 3)
rclmat <- matrix(m, ncol=3, byrow=TRUE)
reclassified <- classify(sthelens_14_65, rclmat, include.lowest=TRUE)

plot(reclassified)
```

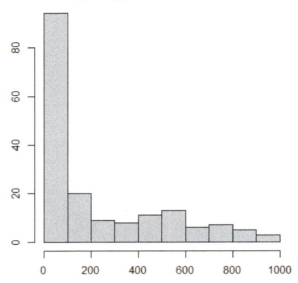

Figure 10.10: This tells us the distribution of the data. Note that there are a lot of values between 0 and 100. This is fairly normal for this type of data, lots of low values, very few high values.

Here we're reclassifying the data as 1, 2 or 3. It is important to note that anything just less than 100 is classified as 1, with 100 being classified at 2. Conventionally classes can be written as:

```
0 - 100 low
100 - 500 medium
500+ high
```

However it might me more accurate to write it as:

```
0 to <100 low
100 >= to <500 medium
500 >= high
```

There are different schools of thought on what is the best way to represent this, and it also depends on what type of data you're working with. With this data, we have decimal values (e.g. 5.59), so we have to be careful with the classes to include all potential values. However, if you were only dealing with integer values, it is easier to work with, because then you can say 0 to 5 is 1, 6 to 10 is 2 and be confident that you are capturing all the possible values.

10.11 Masking

The use of reclassifying data also brings us to the concept of masking. This is where certain areas of a raster are set to 0. This is the area that we want to remove from the dataset, and it is often used to update rectangular raster data to a country outline. We did this earlier when we cropped the age data.

The code we ran has `mask = TRUE` in it:

```
sthelens_0_14 <- crop(age0_14, sthelens_spatVect, mask = TRUE)
```

However, this is not the default. If we ran:

```
sthelens_0_14 <- crop(age0_14, sthelens_spatVect)
```

We would just get a square of data. However with `mask = TRUE` in it the data would be cut to match the boundary itself (see figure 10.11, on the following page):

```
sthelens_0_14 <- crop(age0_14, sthelens_spatVect, mask = TRUE)
```

Behind the scenes, R is converting our vector layer to raster layer, with a 0 outside the boundary and a 1 inside the boundary. Then it multiplies our layer by our mask, to give us the outline we want.

When you use the raster mask to multiply by another raster, any location with a value of 0 will be set to 0 - as anything times 0 is 0.

For hints and tips when (not if) you get stuck, check out Chapter 16: Help! & Where next?

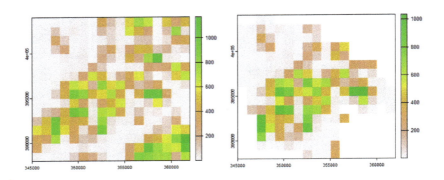

Figure 10.11: Cropping output without a mask (left) and with a mask (right).

Part 4

Project Management

11. File Management

11.1 Learning Outcomes

After reading this chapter, you will:

- Understand why file management is important.
- Know how to use file naming conventions.
- Be able to organise your files in a way that you and others can understand.
- Be able to share your data with a team in a way that works for everybody.

This part of the book contains a number of other topics that, while not directly related to GIS, are really useful and things I think you should know. The first one of these is file management - i.e. when working with GIS how to make sure that you can actually find the GIS files you're looking for. If you look back through the working directories you've created so far, you will see that working with GIS creates lots and lots of files. It is really important to have some type of system to organise these files, so you can find them in the future.

What are files, folders and directories?

If the words "files", "folders" and "directories" mean little to you, the first step is to spend a some time looking through them on your computer. If you're using Windows, you will have a "Documents" folder. Previously, this might have been called "My Documents". If you're using a Mac, you might have a "home" folder.

Open that up and spend a bit of time looking at what is in it. You will probably have some other folders, and a variety of different types of files. It is worth spending a bit of time on this now (30 min to an hour) even if it seems a lot, because you will save much more time than this later on, if you can find the files you're looking for.

There are various tools for navigating and working with files and folders (Windows Explorer on Microsoft Windows, Finder on Mac OSX). Also make sure you understand zip files and how different file types look - e.g. what do shapefiles look like, what do Microsoft Word files look like, and so on. If you want more information, have a look at this video: "GIS File Formats and Good Practice" by Hans van der Kwast, https://www.youtube.com/watch?v=kggwFZHXCl4.

11.2 Working on Your Own

If you're working by yourself, the type of file organisation system you use doesn't really matter, as long as you use a system of some description that allows you to find the files you need to find. You might need to try a few different arrangements and find out what works for you. Here are a few example systems, depending on how complicated you want to make it!

At university, I started with a simple system - a folder for each module I was studying. Anything related to that module went into that folder. I also had an 'Admin' folder for anything else that was university-related, but didn't fit into a module, like timetables. It looked a bit like this (see figure 11.1, on the next page):

Then I could add subfolders as needed. For example, in the Study Skills module we received a PowerPoint file each week, and had to do a short submission. So I had sub-folders (see figure 11.2, on the facing page):

You can have as many subfolders as you like. Use whatever works for

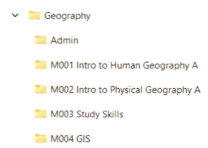

Figure 11.1: An example folder structure, with one folder for each module.

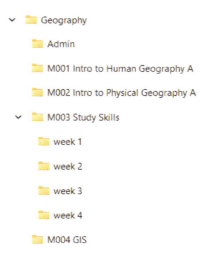

Figure 11.2: Adding in sub-folders for each week.

you. You may have to try some different systems and see what works best for you.

11.3 Working as Part of a Team

If you're working on your own, the system you use isn't crucial. However, if you are working as part of a group, then it is vital you all use the same system. This will make finding files much easier, as well as knowing what the latest version of any file is. It doesn't really matter which approach you use, as long as you all use the same one. It does depend on how many people you have in a project - working with 2 or 3 people is very different to working in a team of 30!

A system which works for larger groups looks a bit like this (see figure 11.3):

Folder	Sub-folder(s)	Description	Type
00_Data management	001_Licences and copyrights	Data licences, copyright statements and the data register should be stored here. A data register is a record or log of all data received, produced and issued to clients (both internal and external).	Read and write
01_Incoming data	[FROM_YYYYMMDD_DESCRIPTION] [FROM_YYYYMMDD_DESCRIPTION] Etc.	Save all incoming data here (i.e. received from external organisations, clients or data downloaded from the internet). Data in this folder should be 'read-only' i.e. it should not be modified, only viewed.	Read only
02_Work in progress	021_Documents 022_Figures 023_MXDs 024_Processing	Used for all work in progress. Store documents (i.e. method statements), figures, map projects (.mxd) and data that you are working on.	Read and write
03_Shared	031_Documents 032_Figures 033_MXDs 034_Processing	Use this folder when you're ready to share your data. Make a copy of it from 02 and save it into the corresponding 03 folder. Data here should be read-only and not modified.	Read only
04_Issued	[TO_YYYYMMDD_DESCRIPTION] [TO_YYYYMMDD_DESCRIPTION] Etc.	Used when you have issued data to a client (internal or external). Read only.	Read only
05_General resources	051_Logos 052_Templates	Used to store logos and map templates (i.e. map projects that have been set up in a particular style).	Read and write

Figure 11.3: A more complex filing system.

Here, this company has a system where they have a different folder for each project. This is then split into the subfolders you see above.

- **00_Data management** contains information on data licenses and copyright statements. It also contains a data register, which is a list of all the data received, produced and issued to clients.
- **01_Incoming data** contains all the data received from outside your group - whether downloaded from the internet or supplied by the client. This is split by folders, identifying where the data came from, when it was received and what it is. This folder should be read-only, i.e. you never make any changes to the data. The last thing you want to do is to have to email your client and say I'm really sorry, please can you send over the data again, as we've lost

it?!

- **02_Work in progress** is a working folder, where you can store your data and files when you are working on them. In this example, they have split them by file type, with MXDs being ArcGIS project files - similar to QGIS .qgz files, or RStudio .rproj files *(see Chapter 13 for information on RStudio projects)*.
- **03_Shared** is for when data is shared internally, so your colleagues can look at it and share comments. This is read only to prevent unexpected changes.
- **04_Issued** is for when data is shared with the client or other users. A snapshot is stored, so you can always check exactly what was shared when, and if the client comes back and says, please can you send over that data again, you can do that quickly and easily. This is read-only, so once you have created the data and sent it off, it doesn't change.
- **05_General resources** is used for logos, templates and other useful files that don't fit elsewhere within the system.

This is quite a complex system, but works really well for projects involving large groups of people. If there are only two or three of you, it might be overkill. However, the key bit is to have a system of some description - and to be consistent in applying that system. You may come up with a different system for your group or organisation - but make sure everyone knows what it is and be consistent when using it!

Along with having a system of folders, it can be very useful to have a system for naming files. Unless you're using a version control system of some sort (see Chapter 12), then you will likely end up with multiple versions of different files. For example, if you're writing a report, it can be easy to end up with something like this (see figure 11.4, on the following page):

A good approach is to use version numbers, e.g.:

- `report-v1.docx`
- `report-v2.docx`

You might end up with something looking like the example from my project folder (see figure 11.5, on page 181).

You could also supplement this with dates if you like:

Figure 11.4: Piled Higher and Deeper by Jorge Cham, www.phdcomics.com

- `report-v2-2021-11-09.docx`
- `report-v2-2021-11-11.docx`

The best format for dates is YYYY-MM-DD, as the computer can order files by name, and in this case they will also be in date order. Using the DD-MM-YYYY format or the MM-DD-YYYY format won't allow you to do this.

If you are working in a group, it may be useful to add initials:

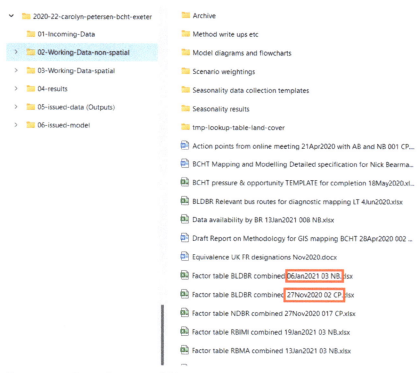

Figure 11.5: Example project folder. I used the system above, but simplified it slightly. We also used a variation of the naming convention with dates, version numbers and initials.

- `report-v2-CP.docx`
- `report-v3-NB.docx`

You can easily see which is the latest report (the highest number), and who was the last person to make changes. My advice would be to never add the word FINAL to a filename - this almost guarantees someone somewhere is going to ask you make some changes to your report, after you name it that!

11.4 Backups

Finally, it is worth spending a little bit of time on backups. I'm sure you've all had backups mentioned to you many times before, but it is well worth repeating.

If you already have some type of backup system in place, you should test it out. Pretend your laptop has been stolen, or completely stopped working. If it helps, ask a friend to physically take away your laptop for a day. Where would you find a copy of your files? How would you carry on working? If you have your files on a cloud service (e.g. Dropbox, Google Drive or OneDrive) or an external drive, have a look at some files - are they up to date? Do they open? How east is it to pick up your work where you left it the previous day?

Testing this out now is crucial - then you can discover if there is a problem with your backup, and solve it now, rather than waiting until your laptop is actually stolen, or actually dies!

If you don't already have a backup system in place, anything is better than nothing. However, a great approach is something called the 3-2-1 backup strategy. This means that you should have 3 different copies of your files, with 2 stored locally on different 2 mediums (i.e. devices) and 1 copy stored off-site. Three sounds like a lot, but it isn't too difficult to manage, and you may be doing some of this already.

For example, if you store your files on your computer, this is 1 copy. This is what we call 'on-site' - i.e. physically with you, wherever you work. If you store your files in OneDrive, Dropbox, Google Drive or something similar then these files are also stored on the cloud - and would count as copy number 2, and are located 'off-site'. So if your home / office burnt down, your computer was stolen, etc. you could still get to your copy of the files in the cloud. However, you're then reliant on Dropbox, OneDrive etc. working all the time and keeping your files safe. Another approach is to regularly copy your files from your computer on to a USB stick or external hard disk. You then have a separate copy of your files (this is copy number 3).

How often you should copy files to your USB or external hard disk is a good question. It depends a bit on how often your files change, and how much work you would have to re-do if your laptop stopped working. There is an element of redundancy here, as the USB stick is not your only copy (you have Dropbox etc. as well). As a starting point, I would suggest updating the copy of your files on USB every month.

There are many different tools out there to help you with this - I use something called FreeFileSync[37], which allows me to compare and synchronises files from my Documents folder to an external hard disk. Whatever you choose to use, choose something!

If your employer / university provides your computer, they may well be making backups for you already. Sometimes they will back up your Documents folder or a specific network drive - find out from them what they do. This will probably be 'off-site' (i.e. a different place to your computer). Also find out how to get hold of them if / when you need to.

Once you've decided how you want to back up your files, make sure you test it out as well. Pretend your laptop has died (get a friend to take it away for a day). How do you find your files? What work can you do? What can't you do? What have you completely lost?

I hope I've managed to convince you of the importance of having a file management system. A little bit of time invested up front will save a lot of stress, heartache and time trying to find files if you don't have a system of some description. Overall, there are no hard and fast rules with this, just three key principles: everyone should use the same system (in a project), never use the word "final" (in a file name), and test your backups!

---

[37]https://freefilesync.org/

12. Version Control using Git & GitHub

12.1 Learning Outcomes

After reading this chapter, you will:

- Understand how we can use version control to help manage R scripts.
- Know how to work with a version control system (Git) and hosting service (GitHub).
- Be able to use GitHub when working with your future self, with a group of collaborators and on open source projects.

Version Control is a really useful concept from computer science that has some great applications when working in R. At its core, version control allows multiple people to work on the same project at the same time. It works best when working in plain text, and if every change someone makes is logged. If necessary, you can then go back to the previous version very easily - without multiple versions of files cluttering up your working directory. Different people can easily work simultaneously in the same project on different files, and potentially even work on the same file at the same time. Everyone's work is then collated together. The use of version control allows this to happen in a structured way, and if two people have made two different changes to the same bit of the same file, this can be identified and managed easily.

There are several different types of version control software - Git is by far the most commonly used, although there are others. Git is what we're focusing on in this book, as it integrates well with RStudio. We could run Git on our own computer for version control, and it would work very well. However, the big benefit of using version control is when we are collaborating with other people.

GitHub is a web-based service that makes this collaboration work. The

GitHub website had a range of additional features that support social collaboration when making changes to code, as well as providing a place to store a copy of the work that others can contribute to.

12.2 Git Basics

Git has a wealth of specific terminology, which at first glance can be a bit intimidating, but the basics are all you need to start.

Repository: A place where you save your code and do version control. Typically, you would have a different repository for each project or thing you are working on. Typically in an R project your working directory would be your repository.

Add: The process of adding files to a commit, i.e. you make some changes to a file, and then save it to your computer, and add it to Git. This is sometimes called adding files to the staging area. You can add multiple files at this point.

Commit: Take all the files you have added to the staging area and save them into the repository. At this point you write a little message saying what you did - this is called a commit message. For example, you might say: "added new file for documentation", "add this line of code", "fix this problem" or whatever change you made. These messages help other people understand what change you made - very similar to writing comments in code!

If you're just working on your own, and saving to your own computer, that is it - you've successfully created a repository. When you make another change to the code, and save it again, Git will save your new version and the old version, so if you ever decide you want to go back to the old version you can. Comparing different versions is called a 'diff'.

Diff: Show the difference between two commits, or files at two different points. Git will show what has been added and what has been removed. You can revert (go back to) earlier versions at any point in time.

However, the real power comes from sharing this with others. The key to this is sharing our repository and allowing others to commit changes to it.

12.3   GitHub Basics

Push: The process of copying commit(s) from your local repository to a remote repository on the internet (e.g. GitHub).

On GitHub, these repositories can either be private, shared with just you or specific other people, or public, where anyone can see them.

Pull: Other users can then Pull the repository, i.e. copy it from the remote service (e.g. GitHub) to their own local machine. They can then make changes themselves, Add and Commit the files, and Push them to GitHub (if they have permissions).

The Add / Commit / Push / Pull commands are all structured so multiple people can use them at the same time. For example, it's easy for Alice to *Pull* the code, and work on file A. Bob can also pull the code and work on file B. Then when each of them *Pushes* their changes, GitHub realises they have been working on different files, and updates the versions on GitHub. Therefore, the copy on GitHub will be updated with file A from Alice and file B from Bob. Their comments allow each other to see what they have been working on. At some point (ideally as soon as possible) Alice will need to make sure she *Pulls* Bobs changes, so she can see them on her computer. The same thing applies for Bob.

Git is designed that these changes are integrated easily. If people are working on different files, that is no problem at all. If Alice and Bob were working on the same file, Git would flag this and say, "Do you want to integrate the changes?" and it would give you a little preview of the changes in the different sections. Often, if Alice and Bob were working on different sections of the same file, Git is clever enough to see this, and can integrate both of their changes into the file without breaking a sweat.

12.4   Pull Requests and Branches

If Alice and Bob were working on the same section of the same file, then Git would flag this and say it couldn't combine them automatically. Someone (either Alice and Bob working together, or their boss,

Charlie) would then have to decide whose changes to keep. Most likely, they would create a third, new version that integrated the best bits of both of their code. This is where some of the features of GitHub come in. If there is a conflict, with Alice and Bob changing the same bit of code, then someone could create a *Pull Request*. This allows multiple people (Alice, Bob, Charlie and anyone else) to comment on the code and have a discussion. They can agree what changes they want to integrate, update the *Pull Request* with the new versions and then and *Merge* the changes in.

You can also have different levels of permission on Git. For example, Bob might only be able to read the code on GitHub, but not write to it. In this case, Bob can still edit the code, but he creates his own copy of the repository on GitHub (a *Fork*) and then Bob pushes his code to that. This doesn't change the main repository, so what he does is create a *Pull Request*, saying "here are some changes I have made, what do you think?" You can then look at that, and say great, accept the *Pull Request*, and it goes into your repository, or send a message saying can you change this (they can submit a new *Commit* and update the *Pull Request*) or you can say no, reject the *Pull Request*. This allows different people to work on the same set of code, and their changes to be integrated in a structured, systematic way.

The other aspects of Git worth mentioning briefly are branches. This is where one repository has multiple different versions of the code - a little like *Pull Requests*. For example, Daniel might create a branch where he is adding in a new feature. He can work on this separately to the main code. Elizabeth might have to make some bug fixes - she can do this on the main branch, separately from Daniel's work. At any point, you can see what changes Elizabeth has made and whether they clash with any changes Daniel has made.

12.5 Sharing Openly

Many open source projects are available on GitHub, and use this process to manage their code and updates to their code. There are also lots of ways of contributing apart from writing code - for example, testing, bug reporting, writing documentation and answering users' questions. If you want to contribute code (or documentation), then you will most likely do that through a pull request. Most projects have guidance for this, so do read this before starting to contribute.

12.6   Working in a Team

You can use the same technology to work on a closed project within a team. Good communication is key here, so everyone knows who does what. Writing comments in your code (1) and writing good commit comments (2) are key here, so everyone knows who has done what, and what the different sections of code do.

This begins to feed into the SCRUM process used by some coding companies, where people report what problem they're working on (repository and/or issues) and what problems they've encountered. GitHub can be used to manage this process, and there are also many other tools for this.

> **Working on your own**
>
> Git can also be a really useful tool for working on your own. Instead of collaborating with other people, you are, in fact, "collaborating with your future self". So when working on a project, using Git allows you to always have a copy of that old version, in a structured way, without having to manage many versions of files. I've used it for writing this book, and for writing many other things. Always having the option to go back to a previous version is really, really useful and gives me confidence that I won't lose any of my writing!

Overall, Git is a great way of working with version control to help you manage plain text files in your project. It is really useful when you're working in a group, or when you're working on your own. There are also lots of extra features you can use if your work needs them. There are quite a number of terms, but I think it is one of those things that makes more sense when you start using it. Most tutorials, including the ones below, include some practical elements, so you can try it out and see how it works. Check out the introduction to Git and the tutorial below.

12.7   Further Git Resources

The Data Carpentry's do a great introduction to Git,[38] and if you want a more workshop based one, I can recommend Archaeogeek's post,[39] which takes about 45 minutes to run through.

---

[38]https://datacarpentry.org/rr-version-control/
[39]http://archaeogeek.github.io/foss4gukdontbeafraid/

13. Projects in RStudio

13.1 Learning Outcomes

After reading this chapter, you will:

- Understand how to use projects in RStudio.
- Be aware of some of the advantages of using projects.
- Understand how RStudio projects relate to version control repositories.

So far we've used RStudio to work with scripts, and we've mentioned working directories and version control. RStudio has the ability to work with projects, which link some of these concepts together. A project in RStudio can store your working directory, your environment and remember which scripts you have open, so when you reopen a project, it will restore your current working state.

It will also integrate a number of other aspects, including version control.

13.2 A New Project in a New Directory

- To start a new project in RStudio, select **File > New Project...**
- Click **New Directory** to start a completely new project. This directory will become your working directory.
- There are a number of templates, **New Project** is a blank template.
- Choose where to save your new project.

RStudio will close any other files you have open and give you a blank working environment.

Try adding a new script and write something in it that will create a new variable.

- Save your script. By default, it will save it in your working directory.

Try closing RStudio. It will probably prompt you, asking if you want to save your workspace image. This is the variables you have listed in the Environment. Click **Save** this time.

Reopen RStudio, and it will automatically reopen the last project you had opened. In this case, it should reload the script you had open, and the variable you created in the environment.

RStudio remembering which scripts you had open is very useful, and this is something I use all the time.

However, there are different schools of thought on whether you should save the Environment or not when using a project. Personally, I would say it is better practice not saving the Environment each time you close a project. This is for a number of reasons:

- When working with spatial data, the file that the environment is saved to (.RData) can get quite large (several megabytes or more) very quickly. Every time you open this project, this has to be loaded, and it can start to become quite slow.
- If you've written your script correctly, it should be able to be re-run every time you open the project, reading in your data again and doing any processing you've done. Doing this every time you open RStudio is a good way of making sure what you think is happening in the code is actually happening.
- If you do save the Environment, it will be reloaded each time you open RStudio. If you forget where you got to in the code, then it may not be clear what the variables in the environment are. If you don't re-run your code from scratch, you might end up with variables that are out of date, wrong, or not what you expect them to be. This can cause your code not to run as you expect it to, and can be quite tricky to diagnose what is wrong.
- Also, the environment files are in binary format, and don't play well with version control.

The exception to this is if you're reading in a large amount of data, which can take quite a long time. By large, I am talking gigabytes of data which can take tens of minutes to read into R. It can be quicker to store the data in the .RData file and read this each time you open

RStudio.

Other people would say that it is easier to keep the environment each time you open the project, as you don't have to rerun all the code to get to where you are.

I would say whichever approach you take, make sure you're happy with what is happening, and make sure you know what is being stored and what is not being stored. This can save confusion later on!

13.3   A New Project in an Existing Directory

If you already have a working directory setup - perhaps with some data files and scripts already in it - we can get RStudio to set up a new project within that directory.

- To start a new project, select **File > New Project...**
- Click **Existing Directory** to start a new project in an existing directory.
- RStudio will ask you to identify the directory you want to use. Do this, and then click **Create Project** to set up the project.

Everything in the previous session about projects still applies - so do also have a read through that!

13.4   A New Project from Version Control

You also have the option of creating a new project by checking it out from a version control repository. RStudio has a very good integration with Git, and you can run a number of Git comments (Add, Commit, Push, etc.) from within RStudio, which can be very useful. Check out Chapter 12 on version control if you haven't read it already.

Following the wizard here will allow you to close a repository and work on it on your own computer.

You can also use other tools to clone repositories to your computer. The Git command line and GitHub Desktop are both useful tools that I use a lot. You can also mix and match - you can set up the repo using GitHub Desktop and then make commits using RStudio - it is all interoperable.

One thing I would recommend is getting into the habit of primarily using one tool for one task. Theoretically, you could use RStudio to add some files to a commit, and then use GitHub Desktop to add more files to the same commit and push it to the repo. However, there is a great risk for confusion there and anything that you can do to make things easier; I would thoroughly recommend taking the easier route!

This chapter has given you a brief overview of RStudio Projects. If you want more details, I would recommend having a read of this article[40] from Posit, the producers of RStudio.

---

[40]https://support.posit.co/hc/en-us/articles/200526207-Using-RStudio-Projects

14. R Markdown

14.1 Learning Outcomes

After reading this chapter, you will:

- Understand what Markdown is and how it is useful.
- Realise why Markdown is useful for Reproducible Research.
- Realise why Markdown is useful for Version Control.
- Understand how Markdown is related to LaTeX.

Alongside version control, another really useful skill when scripting in R is *Markdown*. Markdown allows you to write in plain text, and add specific characters to add styling to the text.

Similarly to an R script, Markdown is written in plain text. This means it works very well with version control, as the concept of writing things in plain text is key for version control to work efficiently. Another advantage is the fact that plain text files are much smaller than alternatives, meaning they're much faster to work with. With Markdown addons (and also LaTeX, see below) you can create very nice looking documents written in Markdown format.

Markdown consists of a series of tags that allow plain text to be converted into nice looking documents. It works a little bit like HTML, and the source is very human-readable. For example:

There are also many other things you can do in Markdown. It is used a lot in (simple) documentation, and GitHub can compile it automatically, allowing you to create a website. A nice example of this is my Installing R & RStudio guide[41] which is created from a markdown page from this[42] GitHub repository.

---

[41] https://nickbearman.github.io/installing-software/r-rstudio
[42] https://github.com/nickbearman/installing-software/blob/master/r-rstudio. md

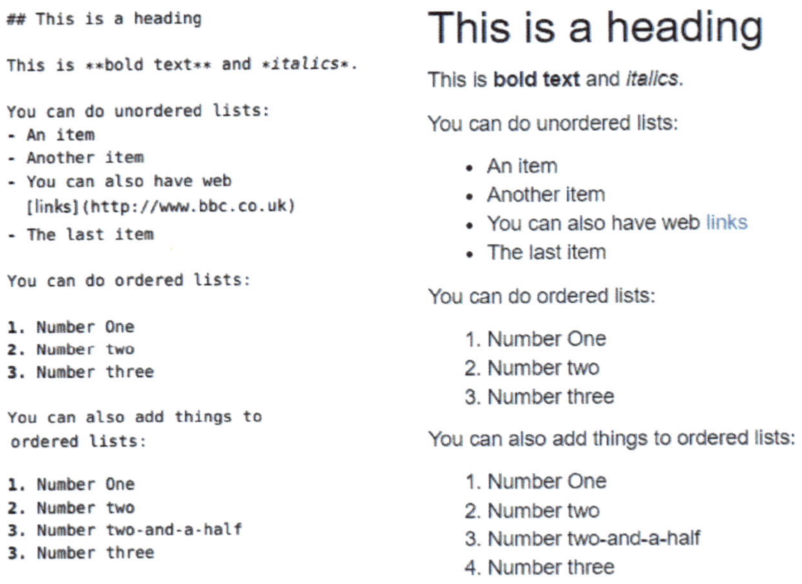

```
## This is a heading

This is **bold text** and *italics*.

You can do unordered lists:
- An item
- Another item
- You can also have web
  [links](http://www.bbc.co.uk)
- The last item

You can do ordered lists:

1. Number One
2. Number two
3. Number three

You can also add things to
ordered lists:

1. Number One
2. Number two
3. Number two-and-a-half
3. Number three
```

This is a heading

This is **bold text** and *italics*.

You can do unordered lists:

- An item
- Another item
- You can also have web links
- The last item

You can do ordered lists:

1. Number One
2. Number two
3. Number three

You can also add things to ordered lists:

1. Number One
2. Number two
3. Number two-and-a-half
4. Number three

Figure 14.1: Markdown code (left) and complied output (right).

When tied-in with version control, anyone can contribute to Markdown documents. For example, if you knew a new version of RStudio had come out, you could update the website for me - by changing the version of R listed in the markdown file, and creating a pull request (see Chapter 12 for more details on version control).

14.2   R Markdown

Where Markdown comes in really useful is a slightly different version, called R Markdown. This allows you to include R code in your document, which can run when you compile the document (i.e. when you turn it from plain text into a PDF). This means you can create dynamic reports, include the data you're analysing in your analysis directly in the document, and allow you to do fully reproducible research.

Reproducible research is a key concept when performing academic work, essentially meaning that anyone with a similar level of knowledge should be able to replicate your work. This means providing the data you used, and the code you used to perform the analysis.

There is a whole area of research on reproducible research, and its implementation in different academic areas. If you're interested in learning more, I would suggest:

- The Turing Way's guide to Reproducible Research[43]
- AGILE, a GIS conference series asks for people submitting their research to be reproducible and have some very useful guidelines[44]
- Many academic journals also request that authors submit reproducible code with their journal article for review.

It is also very useful in training - for instance, for my training courses I have created worksheets as R Markdown files - which makes them easy to update and change if R is updated. As they're on GitHub, anyone else can easily correct or contribute to them as well (see Chapter 12 on Version Control).

14.3 Quarto

Quarto is a development of R Markdown, created by the makers of RStudio. Quarto allows you to create documents with R code in them (exactly like R Markdown) but also with code from other languages in them - for example, Python. This allows you to create documents with multiple programming languages in them, which make it a great tool for learning. Quarto has also been extended to allow you to create websites, presentations and many other outputs with it. See the Quarto website[45] for more details.

14.4 More Advanced Methods

There are also a range of other more advanced methods - Juypter Notebooks, LaTeX and Markdown for Presentations.

---

[43]https://the-turing-way.netlify.app/reproducible-research/reproducible-research.html

[44]https://osf.io/numa5

[45]https://quarto.org/

Jupyter Notebooks are an interactive version of these types of analysis files.[46] You can create them with R and Python, and you can write up your analysis in them, with the advantage that anyone reading them can actually run the code you've written, dynamically, and change it if they want too. There are great learning tools that can be really, really useful.[47]

LaTeX is related to Markdown, in that it is a typesetting language. La-TeX is more complex than Markdown, but also much, much more powerful in that it can do equations, and you can specify many hundreds of different layout parameters (margins, font size, page layout, headers, footers, figure positioning, etc.). Being plain text, you can also use version control with it. (LaTeX resources).

RST (ReStructured Text) is another language, about half way between Markdown and LaTeX in complexity and is, in fact, what this book is written in!

Finally, there are a number of ways of writing presentations (i.e. PowerPoint presentations) using these same Markdown techniques. Currently, the most popular is Reveal.js, which includes a range of different options for making really good-looking presentations. Learn more at the Reveal.js website[48].

LaTeX can be great for avoiding things like this (see figure 14.2):

Figure 14.2: Issues with using Microsoft Word with large documents, from AndrewArruda and k4hsm on Twitter.

Although LaTeX has its downsides, such as being a bit complex to set

---

[46]Information   on   Jupyter   Notebooks: `https://datacarpentry.org/`
`python-ecology-lesson/jupyter_notebooks.html`

[47]Jupyter Notebook and setup instructions: `https://carpentries-incubator.`
`github.io/python-humanities-lesson/`

[48]`https://revealjs.com/`.

up and get working initially, it is well worth spending a bit of time with to see if it works for you.

Overleaf[49] is a web-based tool that allows you to write your document in LaTeX, either in code, or using an editor a bit like Microsoft Word. It's a great way of getting into LaTeX without it being too daunting!

---

[49]https://www.overleaf.com/

Part 5

Next Steps

15. Finding Spatial Data

15.1 Learning Outcomes

After reading this chapter, you will:

- Know how to find different sets of spatial data.
- Understand how to check if a set of spatial data is useful for your work.
- Know that finding the data is only the first step; there are a number of important steps you need to take to make the data useful for your work

This chapter looks at different sources for acquiring spatial data, and how to evaluate whether a specific set of data is appropriate or not.

In any GIS project, data is key. However, finding data can be a real problem - and it is not unusual for the data acquisition element of any GIS project to take up to 50% of the time spent on the whole project. This includes finding the data, getting it into the right format, and evaluating whether it is useful and appropriate for your work. When looking for data, it is important to think about what data you're looking for, and what do you want to be able to do with it. For example, what analysis are you doing, or what map do you want to create?

Once you're happy with what data you're looking for, there are two key steps; firstly, finding the data, and secondly, checking whether the data is appropriate for the analysis you want to do.

15.2 Finding the Data

There is no one repository for spatial data - anyone can make spatial data available, and many different organisations do. Where you look for it depends on what type of data you're looking for - both in terms of the geographic area it covers, and what the data contains (sometimes called the thematic area of the data).

I will include a few specific resources below. However, one of the best lists of GIS data on the internet is Free GIS Data,[50], curated by Dr Robin Wilson, a freelancer in Remote Sensing, GIS, Data Science and Python (see figure 15.1). This is a long list of different, free to use GIS data sets, grouped both by country and by theme. Sometimes you need to register to use the data, but this is always free to do and most of the time automatic so you don't need to wait.

Figure 15.1: The Free GIS Data site, freegisdata.rtwilson.com, a great list of free to use GIS data sets, sorted by theme and country, created by Robin Wilson.

Of particular note is this very important heading Robin has posted on the page:

Beware: The data linked to below may be inaccurate, incomplete, or just plain wrong. As always, critically examine the data you are using, look at what organisation produced it and what agenda they may have, and beware that there are disputes over some of the data (particularly country boundaries).

This is really important and goes for any data you download and use - make sure it is appropriate for what you want to do with this data.

---

[50]https://freegisdata.rtwilson.com/

See the next section (Is the data appropriate for your analysis) for more details.

There are also many other useful sites I can recommend, along with some useful hints and tips.

Ordnance Survey Open Data - osdatahub.os.uk

- Ordnance Survey is the national mapping agency for Great Britain[51] and has a wide range of data available. They have a subset of their data available as Open Data, which will be useful for many projects. I use the OS Open Greenspace data in one of my training courses. Find more details on OS Open Data on the Ordnance Survey website[52].
- If you're working in academia, you may also have access to Digimap, which provides academics with access to more detailed Ordnance Survey data. The main one of these is MasterMap, which is the most detailed data Ordnance Survey provides. Check out their website for more details.
- If you are working in the public sector in the UK, you may have access to more Ordnance Survey data through the Public Sector Geospatial Agreement. See the Public Sector Geospatial Agreement page on the Ordnance Survey website[53] for more details.
- If you are working outside Great Britain, most countries have a national mapping agency that does a similar role to Ordnance Survey. Sometime this is organised at a country level, but other times this is done at a state or regional level. Usually it depends on exactly what type of data you are looking for. Check out the Free GIS Data list[54] for the country you are working in. Not all offer free data - and therefore may not be on the list.

GeoBoundaries - geoboundaries.org

- This website hosts national and sub-national boundaries for every country in the world. Sub-national boundaries go down two or

---

[51]Ordnance Survey Northern Ireland cover Northern Ireland, and have a different range of data available. They are also available through Digimap and have some Open Data at `https://www.nidirect.gov.uk/services/osni-open-data`.

[52]`https://loc8.cc/rgis/osopendata`

[53]`https://loc8.cc/rgis/psga`

[54]`https://freegisdata.rtwilson.com/`

three levels, and use whatever terminology that country uses for them - i.e. states, provinces, counties, municipalities, etc. Are most likely to be up-to-date and I would recommend this over GADM.

GADM - gadm.org

- Global ADMinistrative boundaries - contains national and sub-national boundaries for every country in the world. Sub-national boundaries go down two or three levels, and use whatever terminology that country uses for them - i.e. states, provinces, counties, municipalities, etc. Sometimes this site can be a bit tricky to use, and sometimes the links it provides to the data don't work. GeoBoundaries (see above) is a good alternative to this site.

OpenStreetMap - openstreetmap.org

- Often described as a Wikipedia for maps, which is a very good description of OpenStreetMap. Anyone can contribute data to OpenStreetMap, and the data is freely available for anyone to use. The data that is available is based on what people have added, so there is not 100% coverage. However, most urban areas have very good coverage, and there aren't many areas where there is no data at all [13] and [2]. Whether this is useful depends on what data you're looking for. Data can be extracted as a shape-file in a number of different ways, GeoFabrik (under Export) is the easiest. This allows you to download a shapefile snapshot of OSM data for whichever country or area you like. This is a sub-set of the data available on OSM. An alternative option is to use the QuickOSM plugin in QGIS - this allows you to search for data with specific tags.

Where are the world's boundaries?

With any boundaries, there is no one accepted version of the world's boundaries. Political disputes around borders often make it tricky to decide whose borders are correct. Kashmir is a great example of this - see this write up by the BBC.[a] Crimea is a very topical example - Russia believes it belongs to them, and Ukraine believes it belongs to them.[b]

MapMen also have some great videos on this, looking at international boundaries:
- Bir Tawil - the land that nobody wants
  http://youtube.com/watch?v=J5iJSXaVvao
- Enclaves in India and Bangladesh
  http://youtube.com/watch?v=r-alzkvPwFo
- How do you start a new country?
  http://youtube.com/watch?v=hX4s1ZLW_PI
- Who owns the South China Sea?
  http://youtube.com/watch?v=ypMCK7NcbxU
- Who Owns Antarctica?
  http://youtube.com/watch?v=Eg1ScKoBnHA

So, how do you decide which borders to show? Sometimes this is decided for you by the source. GeoBoundaries have individual country files (as each country would define themselves) and a Comprehensive Global Administrative Zones data set - Seamless global layers with demarcations for disputed areas. Here they follow the US guidance: "disputed areas are removed and replaced with polygons following US Department of State definitions".

MapTiler gives you the choice - and you can visualise your web map from different countries point of view.[c]

This can manifest itself in all sorts of interesting ways, including on Christmas decorations.[d]

Whatever you decided to do, it is important to be aware of potential issues when showing country boundaries.

---

[a]https://www.bbc.co.uk/news/world-south-asia-11693674
[b]https://www.britannica.com/place/Crimea
[c]https://documentation.maptiler.com/hc/en-us/articles/4405462162065-Disputed-borders-on-your-maps
[d]https://jcheshire.com/christmas/the-contested-borders-on-my-christmas-tree/

Overture Maps - overturemaps.org

- Overture is a new development aimed at providing free geospatial data for developers. It is a combination of many different data sources, with OpenStreetMap as its main source, and other sources used to fill in the gaps, such as Microsoft's buildings data. It is still in its early stages, and accessing the data is designed for developers. The process is outlined in the docs[55] and on GitHub[56] if you're interested. At this point (2024) it's probably not worth spending the time on unless you have a specific reason to use this data source. In the future, it might become more accessible so is worth keeping an eye on.

EuroStat - ec.europa.eu/eurostat

- A useful site if you're after data at the European level. It collates a whole range of Census like statistics and shares then across the European Union at a variety of levels, including Country, NUTS1, NUTS2 and NUTS3 level. A nice example is this map of population.[57]

WorldPop - worldpop.org

- Developed by academics at the University of Southampton, UK, WorldPop provides population data and population projections for most developing countries across the world. They have a variety of different formats, focusing on raster data at 100m resolution, and have yearly estimates for 2000 - 2020. The data are mostly modelled (based on Census data and a number of other sources), and look at worldpop.org for more details on this.

Finally, you can also do a web search for any GIS data you want. Search terms such as 'GIS data' or 'spatial data' or 'shapefile' combined with whatever data you're looking for usually produce some results. There is a skill to searching the internet, and it might take a bit of practice to find what you are looking for.

Alternatively, if you're developing someone else's work, or replicating an existing study, any academic paper should reference where they get their spatial data from, in exactly the same manner as referencing an academic concept or an academic sources. If an article you're reading

---

[55] https://docs.overturemaps.org

[56] https://github.com/OvertureMaps/data

[57] https://ec.europa.eu/eurostat/web/interactive-publications/regions-2023

doesn't reference where they got their data from, I would say it is perfectly reasonable to email the author and ask where they got their data from.

Finding data is a bit of an art, and can take some time, so be patient. Even experts still have to go through this process, find the data, and evaluate whether it is useful for the project or not!

Whenever you have some data, you need to evaluate whether it is appropriate for your project.

15.3   Is the Data Appropriate?

When spatial data is collected, whoever is collecting it is usually collecting it for a specific purpose. This means that the data may or may not be useful for what you intend to use it for. This is when we have to think about whether the data is appropriate for the work you want to do with it.

By 'appropriate' I mean:

- Does the data cover the right geographic location?
- Does it contain the data you actually need?
- Is it recent enough?
- What projection and coordinate system are the data in?
- Is it at the correct scale/resolution (for the analysis/map you want to do)?
- How much data cleaning are you likely to need to do?

This is a short check list, and I will go through each in more detail below.

Location is key to spatial data, and whether the data covers the correct location is very important. Most spatial data only covers a certain area, so you need to ensure that the coverage includes the area(s) you need.

The contents of the data are also important, in terms of what columns are in the attribute data. For example, if you have data on housing tenure, do the attribute columns have the information you need? You might be interested in private rental and social rental housing. However, if the columns only split data by 'owner-occupied' and 'rented', the data won't be any use to use as they do not distinguish between private rented and socially rented.

You should also consider whether the data is recent enough. It is important to note when the data was collected, and whether much has changed since that date. There is no set time period that is 'enough'. These are really dependent on what the data is, as for example, railway line data don't change that frequently, so a 10-year-old dataset might be fine. Whereas a housing data set for an urban area with several new housing estates on it might be very out of date after 10 years. Use your knowledge of the local area to know whether there have been any relevant changes - or if you aren't familiar with the local area, then ask someone who is.

You also need to know what projection and coordinate system the data is in - have a look at Chapter 2 for more details on this.

Also, what scale (or resolution) is the data at, and is this suitable for the analysis you plan to perform? See Chapter 2 for details on this.

Finally, once you've actually found the data, you might need to do some data cleaning. This could be removing outlining points, or incorrectly located points, removing unnecessary columns or generalising or clipping data, so it is easier to process.

Whilst this 'data decontamination' or 'data cleaning' (as it is sometimes termed) can be quite time-consuming, it is worthwhile - because it will allow you to complete your analysis and be confident of the output you've created. Sometimes, in the words of 'data science', this is called 'data wrangling' - which we cover in Chapter 10.

16. Help! & Where Next?

16.1 Learning Outcomes

After reading this chapter, you will:

- Understand where to get help when get stuck with R.
- Know how to post a good question that is likely to get a response on StackExchange.
- Be able to find out about new R libraries which might be useful for you.
- Be aware of other books and resources for performing spatial analysis work in R.

We've covered a lot of the basic skills for using R as a GIS in this book. However, there are many more types of analysis you can do in R, so where do you go from here? Also, I guarantee that all of you will get stuck at some point, so how can you solve those pesky problems with R? The skill is not about avoiding the problems in the first place (as that is impossible) but knowing what to do when you find a problem. We will cover all of these in this chapter.

16.2 Help! Problem solving

Throughout this book we've covered some of the common error messages you might come across while using R. There is a whole list at the back of the book (before the glossary), and I will have an online version that I will update as well - check it out for more details.

Libraries can be particularly tricky, especially if you are updating an R installation (rather than installing from new). I'd recommend having a look at my online guide for troubleshooting R & RStudio[58] for some things to try.

---

[58]https://nickbearman.github.io/installing-software/

If you get stuck with R, the first thing to do is to read the error message. The second thing is then to read the error message again - many of these messages are quite impenetrable, so they can take a little bit of digesting!

Very common ones include missing), missing , or missing ". R is very picky about all of these characters, so make sure these are all there.

If you get a slightly more specific error message, then a web search can be really useful. Search for the error message, and you will most likely get some results on the Stack Exchange website. This is a really useful site where people can ask questions, and other people answer them. There are lots of different areas, and the ones you will probably find are the GIS or the programming Stack Exchange. I would say that 90% of the problems you have in R someone else will have had already, and they will have posted this on a Stack Exchange site.

If you're lucky, there will be an answer there already that works for you. If it does, make sure you sign in and up-vote the answer. If you're unlucky, you might have to do some more digging to find the answer you're looking for. With R, there are a lot of variables that could be causing an issue.

A good approach is to ensure that you're using the latest version of any package that you're using. Run install.packages() again for any package you want to upgrade, and R will reinstall it and upgrade if possible.

If you can't find an answer to your question, you can post one. However, there are a few rules to follow to make sure you get the best answer possible. Firstly, provide some reproducible code and, if needed, data. This will allow anyone to replicate your issue. If they can't replicate it, they can't answer your question. Be clear on what you think should happen - it might be you can do your analysis using a slightly different tool.

Check out this video[59] from a recent FOSS4G:UK conference - How to get a good response on Stack Exchange by Ian Turton.[60]

---

r-rstudio-library-troubleshooting.html

[59]Video: https://loc8.cc/rgis/ianvideo

[60]Slides: https://www.ianturton.com/talks/foss4g2022/slides.pdf

You are also welcome to ask questions on the book website - but you might have a longer wait than on Stack Exchange!

Finally, if you're working in a team or with other people who're doing work in R or GIS, ask them. Your colleagues can be really useful, and sometimes a fresh pair of eyes can see an answer straight away which you couldn't.

Additionally, if you've been doing a lot of coding, and can't seem to get a specific line working, take a break. Programming can be quite intensive, so if you're stuck, don't be afraid to stop, have a break and come back tomorrow.

As a PhD student, I did quite a bit of coding in VBA and JavaScript. Often I'd get stuck on a particular issue and could easily spend two or three hours making no progress at all. Then I would go home and come back in the following day. With a fresh pair of eyes, I could often solve the issue in 20 minutes! So remember - if you do get stuck, have a break.

16.3   New Libraries

One of the big benefits of R is the number of libraries available - and this is increasing all the time. Keep an eye out for new libraries and code that might be useful to you. Increasingly, academic papers are referencing libraries and providing the code they used in their analysis. You can take and reuse this code in your work - providing suitable references, of course.

16.4   tmap v4

At the time of writing, there is a new version of the `tmap` library in development (version 4). This will result in some quite big changes to the `tmap` code used to create maps in R. I will update the code on the website when `tmap` v4 is generally available, so keep your eyes open! Check out section 4.10 (tmap v4) on page 87 for more details.

16.5 dplyr & ggplot

We have mentioned a little bit about the `dplyr` and the wider `tidyverse` already, but it is worth mentioning here with some links. With the development of the `sf` library (see Chapter 7) we can now make use of many of the tidyverse approaches to data analysis (e.g. pipes `|>` or `%>%`) with spatial data. This allows you to apply (nearly) all of the tidyverse material you might come across to spatial data and work with it in the same way. Geocomputation with R uses the tidyverse quite a bit, and there is also a nice set of code intro.[61]

16.6 Other Books

If you want to build on what you've learnt in this book, there are a couple of other books worth looking at.

Geocomputation with R by Robin Lovelace, Jakub Nowosad and Jannes Muenchow

This is a more advanced book, which gets into the more technical aspects of GIS within R, including detailed spatial data operations and geometry operations. It also covers some spatial statistics and applications of R & GIS in transportation, geomarketing and ecology. It is available online on Robin Lovelace's website[62] and as a print book.

An Introduction to R for Spatial Analysis and Mapping by Chris Brunsdon and Lex Comber

This book covers similar material to Using R as a GIS, but with different examples and is at a slightly more advanced academic level. If you would like some different examples to the ones in this book, then this is a great resource. It is available on Sage[63] and Amazon[64].

---

[61] https://mhallwor.github.io/_pages/Tidyverse_intro

[62] https://geocompr.robinlovelace.net/

[63] https://loc8.cc/rgis/introrsage

[64] https://www.amazon.co.uk/Introduction-Spatial-Analysis-Mapping/dp/1446272958

16.7 Python

Another common programming language that you can use for spatial analysis, which I'm sure you've heard of, is Python. In some ways, Python is very similar to R for spatial data: there are a number of different spatial libraries available, and it is quite a flexible language to use for spatial analysis.

In my opinion, R has some clear advantages over Python. First of all, I think R is easier to learn for people completely new to scripting or programming. R less picky than Python about how you write it, and it is easier to just pick up R and run with it compared to Python. R also probably has a wider user base, in terms of people who're experts in other areas using R as a GIS. R is also much easier to install and get going with on a computer - you just install R, RStudio, a couple of libraries, and you're good to go. Also thanks to the RStudio company (now Posit) you can also do all of this in the cloud very easily - go to posit.cloud , register, and you can have a new RStudio project running in your web browser in minutes.

However, Python has some advantages over R as well. Firstly, it is probably a more powerful language than R, and can manage large data a bit more easily. With Python being more of a programming language than a scripting language, it is a bit stricter in terms of how you write it, but once you master that you can do more with it. It is also probably a bit more common in what we might call the 'hard core' data science community. By some, R is seen as a bit light in underpowered, whereas Python can do a lot more. Python also integrates with more programs compared to R - for example, both QGIS and ArcGIS can have plugins and extensions written in Python.

If you want to learn a bit of Python, I would recommend Dani Arribas-Bel's Geographic Data Science course[65]. It's completely open source and takes you through from beginner. It is a whole semester long module, so there is quite a bit of material there. The course has also been developed into a book[66].

---

[65]https://darribas.org/gds_course
[66]https://geographicdata.science/book/intro.html

16.8   Training Courses

Different people learn in different ways, and it might be that you prefer a training course approach rather than book. I know my discipline to work through material on my own can be quite lacking at times! There are a number of courses available on Using R as a GIS, including one run by me. Check out the details on my website[67].

If you want to start building beyond the material in this book, the courses tend to get more specific depending on what you want to learn. For example, courses on using R for agent based modelling or time series analysis are run by the University of Leeds. Meanwhile, the University of Liverpool runs courses on using R for retail modelling, and University College London runs courses on GIS Big Data Modelling.

To find these courses, the internet and your colleagues are great resources - find out what other people have done and what works for them.

16.9   Thanks and Good Luck!

I hope you have found this book useful. Please do let me know what you think - you can leave reviews / comments at:

- Twitter/X @nickbearmanuk
- Mastodon @nickbearman@fossfordon.org
- LinkedIn linkedin.com/in/nickbearman
- Email nick@nickbearman.com

---

[67]https://www.nickbearman.com

Glossary: Using R as a GIS

Terms in *italics* are defined in the glossary,
terms with **brackets()** are R functions

Used to precede a comment: #this is a comment

**$** Used to refer to columns within a data frame: dataframe$column

? shows the help file for that command ?help or ?head

?? search through the help files for any reference to the word you type: ??dataframe

[,] square brackets are used to refer to specific elements in a list or data frame. pop2011[1,] will show the first row and pop2011[,1] will show the first column

{} Used in a **for** loop or a **if** statement

<- assigns a value or output from a function to a variable

abline() adds vertical lines to a histogram, used to show *classification* breaks: abline(v = breaks$brks, col = "red")

Acknowledgements Information required on any map, including copyright or data sources (e.g. for OpenStreetMap or Ordnance Survey Open Data)

aggregate() group the specified data set by a column, applying a function to the values: aggregate(x = LSOA_crimes, by = list(LSOA_crimes$lsoa21cd), FUN = length)

BeiDou Chinese version of *GPS* (see also *GNSS*)

BNG (British National Grid) A *projected coordinate system* used to represent locations in Great Britain, consisting of *eastings* and *northings*, e.g. 603125, 112589 (see also *UTM* and *WGS1984*)

c(,) used to create a list, either numbers c(1,2,3) or *strings* (text) c("Thomas","Richard","Harriet")

Categorical A variable that has a series of values with no inherent order, e.g. country names, also known as *nominal* (see also *variable type*)

Choropleth A type of mapping where different colours are used to represent difference values; can use *categorical* and *ordinal* data

Classes The groups data are put into for a *choropleth* map

Classification How data are classified into different *classes* for a *choropleth* map (see also *jenks*, *equal count*, *equal interval* and *standard deviation*)

classIntervals() calculate the class invervals for the specified data:

`classIntervals(var, n = 6, style = "fisher")`

colnames() shows the names and numbers of the columns in the specified data set: `colnames(hp.data)`

Coordinates The numbers representing a specific location, usually presented in pairs (see also *latitude, longitude, WGS1984, BNG* and *UTM*)

CRS (Coordinate Reference System) The type of coordinates that are used to represent a specific location (see also *WGS1984, BNG* and *projection*)

Correlation A measure of how much two variables are related, measured using a R^2 value

CSV (Comma Separated Values) A standard format of *tabular data*, can be opened in Excel

CSVT An optional file for use with *CSV* files which specifies the *variable type* of each column

Data type How data is stored within the *Attribute table*, can be *integer* (whole numbers), *real* (decimal numbers) and *string* (text)

DEM (Digital Elevation Model) a *raster* representation of the height of the earth's surface

display.brewer.all() shows all the potential colour pallets from the `RColorBrew` library

download.file() download a file from the specified URL

Eastings A *coordinate* that specifies the distance east, in meters, from the coordinates 0, 0 south-west of the Isles of Scilly (see also *BNG* and *northings*)

EPSG code A 4 or 5 digit code used by GIS to define what *CRS* (Coordinate Reference System) a data set is stored in, for example *WGS1984* is 4326 and *BNG* is 27700)

Equal count (Quantile) *Classification* method where data are split into a number of groups by putting the same number of data items into each group, also known as *quantile*, see also *classification*

Equal interval *Classification* method where data are split into *classes* that are evenly distributed, e.g. 0-20%, 20-40%, etc., see also *classification*

F / FALSE used to specify we don't want something, `frame = F` means we don't want a frame

Field calculator Used to calculate new values (e.g. differences) from existing values for all rows in a vector layer, accessed from the *Attribute table*

Fisher *Classification* method very similar to *Jenks*

for () {} begins a loop to make R repeat a command a set number of times: `for (i in 1:length(mapvariables))`

Galileo European Union version of *GPS* (see also *GNSS*)

Geographic coordinate system A coordinate system covering the whole world, usually using degrees (see also *WGS1984, latitude, longitude*)

Geographical Information Science (GIS) The development of the tools, software and processes used in *Geographical Information Systems*

Geographical Information Systems (GIS) Using spatial data to answer questions about our world (see also *Geographical Information Science*)

GeoJSON Vector spatial data file, consisting of *points, lines* and *polygons*; all saved in one file

Geopackage (GPKG) Open format for storing geospatial data, can store multiple *layers* within one file, both *vector* and *raster*, sometimes seen as a replacement for *shapefiles*

GLONASS Russian version of *GPS* (see also *GNSS*)

GNSS (Global Navigation Satellite System) formal generic term for satellite location systems, see also *GPS, GLONASS, Galileo* and *BeiDou*

GPS (Global Positioning System) a series of 24 satellites in orbit around the earth which allow a GPS device to locate itself, with an accuracy of 1m to 10m (see also *GNSS, GLONASS, Galileo* and *BeiDou*)

group_by() used to group data by a specific value

head() used to show the first six rows of the data frame: `head(hp.data)`

hist() shows a histogram of the specified data: `hist(var)`

if () {} begins an if loop to make R choose a specific command based on a cirteria

Inset Map A small map included on the main map to aid orientation, e.g. a map of Ghana might include an *inset map* of Africa to show where Ghana is

install.packages() Install (download from the internet) new *libraries* in R, you only need to do this once, remember to include the quotes `install.packages("sf")`

Integer A whole number used to represent data, can be used in a *choropleth* map (see also *data type*)

Jenks (natural breaks) *Classification* method based on the Jenks algorithm which groups similar data values together, also known as *natural breaks*, see also *classification*

Joining The process of linking attribute information to spatial data, often used so the information can be shown on a *choropleth* map

Latitude A *coordinate* that specifies the distance north or south, ranging from 0° at the Equator to 90° (North or South) at the poles (see also *WGS1984* and *longitude*)

Layers When you add data into a GIS each different file appears as a different *layer*; this allows different datasets to be overlaid on one another (see also *Contents*)

Layout The term for where you create your final map, where you can add maps, *legends*, *scale bars*, *north arrows*, etc. (also known as *Print Layout*, for QGIS)

Legend An important part of any map, showing what the symbols or colours used on the map represent

library() load the specified library: `library(sf)` you need to do this everytime you start *RStudio*

Libraries (or packages) A collection of R functions brought together to extend R's functionality, we use the `sf` and `tmap` libraries, installed with `install.packages("")` and loaded with `library()`

Lines Used in *vector* data sets to indicate a linear feature, such as rivers, roads or railways; is a series of *points* joined together in a certain order

Longitude A *coordinate* that specifies the distance east or west, ranging from 0° at the Prime Meridian (Greenwich, London, UK) to 180° around the international data line (see also *WGS1984* and *latitude*)

MapInfo A commercial GIS software, created by Pitney Bowes, now developed by Precisely

mean() calculate the mean of the specified set of values: `mean(house.prices)`

merge() joins two data frames together using a specified attribute or ID: `merge(LSOA, pop2021, by.x="lsoa21cd", by.y="geography.code")`

Natural breaks (Jenks / Fisher) *Classification* method based on the *Jenks* algorithm which groups similar data values together, see also *classification*

Nominal A variable that has a series of values with no inherent order, e.g. country names, also known as *categorical* (see also *variable type*, *ordinal* and *quantitative*)

North arrow Used to show the direction of North on a map, used to aid orientation (see also *inset map*)

Northings A coordinate that specifies the distance north, in meters, from the *coordinates* 0, 0 south-west of the Isles of Scilly (see also *BNG* and *east-*

ings)

Ordinal Similar to a categorical variable, but with a clear order, e.g. high priority, medium priority, and low priority (see also *variable type, quantitative*)

Pixel An individual unit in a *raster* data set, the size of the *resolution* squared (i.e. for a 100m resolution *raster* data set, each *pixel* would be 100m x 100m, covering 10,000 square meters (or 1 hectare) of land)

plot() Base function in R used for plotting graphs and sometimes raster data

Points A *vector* data type used to indicate a specific location, such as sample collection points, bird nest sites, towns or cities

Polygons A *vector* data type used to indicate areas, e.g. land parcels, counties and fields; is a series of *points* joined in a certain order with the last point linked back to the first point to indicate an area

print() Command used to output some information to the console

Print Layout The term for where you create your final map (see also as *Layout*)

project file A file created by GIS to store the symbology, classification, layouts and a list of the data layers you are using. Project files do not store any spatial data. Examples include .qgz / .qgs (QGIS) and .arpx (ArcGIS Pro)

Projected coordinate system A coordinate system covering a specific, small area, usually uses meters and can me used for measuring distances (see also *BNG, easting, northing* and *UTM*)

Projection The way the sphere shaped earth is distorted to fit on a flat piece of paper (see also *WGS1984, BNG* and *coordinate system*)

QGIS An open source GIS free for anyone to download, use and improve

QGIS project file (.qgz / .qgs) (QGIS) A project file for *QGIS* which contains links to all the data files (such as *shapefiles* and/or *GeoJSON* files) and information on how they are symbolised; the *project file* does not contain the data itself

qtm() a quick way of showing spatial data, either just the data itself, qtm(LSOA) or a quick choropleth map qtm(LSOA_crimes_aggregated, fill = "count of crimes")

Quantile (equal count) *Classification* method where data are split into a number of groups by putting the same number of data items into each group, also known as *equal count*, see also *classification*

Quantitative A numeric variable with an inherent order, e.g. GDP per capita, (see also *variable type*)

R Scripting language designed initially for statistics, expanded with *libraries* to include spatial analysis, usually used through *RStudio*

R² The *correlation* coefficient of two different data sets, a value of 1 is a strong positive *correlation*, -1 is a strong negative *correlation*

Raster A type of spatial data used with GIS, consisting of a regular grid of points spaced at a set distance (the *resolution*); often used to represent heights (DEM) or temperature data (see also *vector*)

Raster calculator Used with *raster* data to calculate differences (subtract) or calculate other indices (e.g. NDVI)

read.csv() read in a *csv* file from the specified path: `read.csv("police-uk-2020-04-merseyside.csv")`

Real A decimal number used to represent data, can be used in a *choropleth* map (see also *data type*)

res() command to display the *resolution* of a *raster* data set

Resolution The size of each *pixel* in a *raster* data set (e.g. 100 meters, 1km, 100km)

RStudio easy to use interface for *R*, allows easy access to *variables*, plot window and use of *scripts*, known as an IDE (integrated development environment)

Sat-nav A navigation system in cars, which uses *GPS* to direct the driver to their destination

Scale The ratio of units of distance on the map to units of distance in the real world; for example 1:25,000 means that 1cm on the map represents 25,000cm (or 250m) in the real world; usually shown on a *scale bar*

Scale bar Used to show the *scale* of a map

Scripts a series of *R* commands that can be run either individually or all together at once

setwd() set the working directory to the specified folder: `setwd("C:/Users/nick/Documents/GIS")`

Shapefile A type vector of spatial data file, consisting of one of *points*, *lines* or *polygons*; represented in *GIS* as one file but in fact consisting of multiple files (between 4 and 6 files, with extensions of .shp, .shx, .dbf and .prj)

spatial autocorrelation A measure of how much the location of spatial data is correlated with itself, a value of 1 is a strong positive autocorrelation meaning data are spatially clustered with low values near each other and high values near each other, 0 is no spatial autocorrelation meaning data are distirbuted at random and space is not important in this data set, -1 is a strong negative correlation with low and high values as mixed as possible

st_as_sf() use to create a spatial data frame from a non spatial data frame: st_as_sf(crimes, coords = c('Longitude', 'Latitude'), crs = 4326)

st_join() spatial join, joining data by location: st_join(LSOA, crimes_sf_bng)

st_read() read in a shapefile or other spatial data file: st_read("england_lsoa_2021.shp")

st_transform() reproject the specified spatial data object to the specified *crs*: st_transform(crimes_sf, crs = 27700)

st_write() save the specified spatial data object, e.g. as a shapefile: st_write(LSOA_crimes_aggregated, "LSOA-crime-count.shp")

Standard deviation *Classification* method based on standard deviation and mean of the data set

String A piece of text (e.g. a name) used to represent data, cannot be used to create a *choropleth* map (see also *data type*, *real* and *integer*

Style (QGIS) / **Symbology** (ArcGIS Pro) The options to choose the colours and/or symbols to represent data on the map; accessed through right-clicking on the layer and selecting properties and navigating to the Style tab)

style (R) Parameter used to specify which classification method to use for a choropleth map: style = "equal"

T / TRUE used to specify that we do want something: legend.hist = T means we want to show the legend histogram

table() count how many different values there are in the specificed column, and output as a table: table(crimes_sf_bng$Crime.type)

Tabular data Data laid out in rows and columns, as used in Excel (see also *CSV*)

title.position specify the location of the title on a map: *title.position* = c(0.7, "top")

tm_borders() specify the colour and thickness of a border around a map

tm_compass() specify the location of the north arrow or compass on a map: tm_compass(position = c(0.3, 0.07))

tm_dots() specify the symbology of a point layer on a map: tm_dots(size = 0.1, shape = 19, col = "darkred", alpha = 0.5)

tm_layout() specify a range of options in the map layout: tm_layout(frame = F, title = "Liverpool", title.size = 2)

tm_lines() specify the symbology and options of a line layer on a map: `tm_lines(col = "black")`

tm_polygons() specify the symbology and options of a polygon layer on a map: `tm_polygons("Age00to04", title = "Aged 0 to 4", palette = "Greens", n = 6, style = "jenks")`

tm_raster() specify and plot a raster data set: `tm_raster(palette = "Greens")`

tm_scale_bar() specify the location and style of the scale bar on a map `tm_scale_bar(width = 0.22, position = c(0.05, 0.18))`

tm_shape() specify the layer shown on a map, often used in conjunction with *tm_polygons()*: `tm_shape(LSOA)`

tmap_mode() change the tmap mode from plot (default) to view (slippy map with basemap) or back: `tmap_mode("view")`, `tmap_mode("plot")`

tmap_save() save the map to a specified output file: `tmap_save(m, filename = paste0("map-",mapvariables[i],".png"))`

unzip() unzip a file: `unzip("sthelens.zip")`

UTM (Universal Transverse Mercator) A type of *projected coordinate system* used to represent any location in the world, consisting of a series of zones and a set of *coordinates* for each zone, in meters (see also *BNG* and *WGS1984*)

Variable type Information on the type of information within a variable, can be *categorical, ordinal* or *nominal*

Vector A type of spatial data used with *GIS*, consisting of *points, lines* and *polygons* (see also *raster*)

Vertex (vertices) Name for each of the points that connect the *line* segments of a *line* or *polygon shapefile*

View() view the specified data in a new tab in RStudio: `View(hp.data)`

WGS1984 A *coordinate system* used to represent any location in the world, consisting of *latitude* and *longitude* e.g. 51.0426 N, 1.3772 E or 51° 2' 33.53" N, 1° 22' 38.23" E (see also *BNG* and *UTM*)

which() function to select which elements or rows match the specified criteria: `which(tram_stations$RSTNAM == "Chorlton")`

Bibliography

[1] N. Bearman. GIS: Research Methods. Available at: https://www.bloomsbury.com/uk/gis-9781350129559/, 2021. Accessed: 31 May 2021.

[2] Filip Biljecki, Yoong Shin Chow, and Kay Lee. Quality of crowd-sourced geospatial building information: A global assessment of openstreetmap attributes. *Building and Environment*, 237:110295, 2023.

[3] C. Brunsdon and A. Comber. Opening practice: supporting reproducibility and critical spatial data science. *Journal of Geographical Systems*, 23(4):477–496, 2021. doi:10.1007/s10109-020-00334-2.

[4] J.F. Claerbout and M. Karrenbach. Electronic documents give reproducible research a new meaning. In *SEG Technical Program Expanded Abstracts 1992*, pages 601—604. Society of Exploration Geophysicists (SEG Technical Program Expanded Abstracts), 1992. doi:10.1190/1.1822162.

[5] G Darkes and M. Spence. *Cartography: an introduction*. British Cartographic Society, London, UK, 2017.

[6] Shunlin Liang and Jindi Wang. Chapter 4 - atmospheric correction of optical imagery. In Shunlin Liang and Jindi Wang, editors, *Advanced Remote Sensing*, pages 131–156. Academic Press, second edition, 2020.

[7] T. M. Lillesand and Kiefer R. W. *Remote Sensing and Image Interpretation*. Wiley, New York, USA, 2000.

[8] C.D. Lloyd, G. Catney, P. Williamson, and N. Bearman. Exploring the utility of grids for analysing long term population change. *Computers, Environment and Urban Systems*, 66:1–12, 2017.

[9] J. Lovelace, R. Nowosad and J. Muenchow. *Geocomputation with R*. Routledge, London, UK, 2019.

[10] D.T. Lykken. Statistical significance in psychological research. *Psychological Bulletin*, 70(3):151—159, 1968.

[11] D. H. Maling. *Coordinate Systems and Map Projections*. Pergamon Press, Oxford, UK; New York, USA, 1992.

[12] G. A. Miller. The magical number seven, plus or minus two: Some limits on our capacity for processing information. *Psychological Review*, 63:81–97, 1956.

[13] M. Minghini and F. Frassinelli. Openstreetmap history for intrinsic quality assessment: Is osm up-to-date? *Open Geospatial Data, Software and Standards*, 4(9), 2019. doi:10.1186/s40965-019-0067-x.

About Locate Press

Locate Press is a book publisher, focusing on the open source geospatial niche. Many traditional publishers see geospatial books as either scientific content or geared primarily toward consumers. Unfortunately, this means they don't give them the long term care they truly deserve. With more and more technical users using open source geospatial technology (for a wide variety of reasons!), now, more than ever, you need comprehensive and reliable education and training resources.

You've come to the right place!

We know that niche is not a swear word, but a marketplace that needs serious support. Geospatial data management is a core technology for government and business, making practical teaching materials for industry and higher education crucial. We also know that reliable availability of material is key. Our books, once available, remain available long after the first few thousand are sold so that you can depend on them for course material and reference long into the future.

If you are an educator looking for high quality curriculum, we would like to hear from you. Aside from training books, we also aim to provide workshop guides and exercise booklets that you can use in your courses!

Academia is not the only place for learning and training, and Locate Press supports consultants delivering workshops and seminars. If you have solid, practical material that needs some professional polish, give us a call. Likewise, if you need bulk orders to serve your students, or to resell, we can help there too.

Locate Press was founded by Tyler Mitchell in 2012, it's flagship book being *The Geospatial Desktop* by Gary Sherman. In 2013 Gary Sherman took over the helm as publisher, guiding the company until 2021 when Locate Press returned to Tyler.

ORDER DIRECT AND SAVE UP TO 30%

Our paperbacks can be ordered directly from with bulk discounts on 5, 10, and 25+ units: store.locatepress.com

We print in countries that are closest to you and can deliver almost anywhere. Our print books also sell through Amazon or Ingram.

Ebooks (PDF) are ordered and downloaded directly from locatepress.com/ebooks

Educators contact us for desk/review copies:
(250) 303-1831 or
tyler@locatepress.com

Writing for Locate Press

Are you passionate about open source software? Have an uncontrollable urge to share your knowledge with the world?

At Locate Press we're looking for books that open up the world of geospatial. We love concise, targeted titles that help people expand their knowledge and get up to speed quickly. That being said, we don't go around with blinders on—we're open to other leading edge topics related to open source.

We help put your ideas into book form, getting your expertise on paper and in print. Don't let the process scare you, we're here to guide and help all along the way—from outline to print-ready copy.

Le guide du programmeur PyQGIS
Gary Sherman, Noureddine Farah

Spatial SQL
Matthew Forrest

How to Succeed as a GIS Rebel
Mark Seibel

Earth Engine & Geemap
Qiusheng Wu

Using R as a GIS
Dr. Nick Bearman

Discover QGIS 3.x - 2nd Edition
Kurt Menke

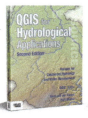

QGIS for Hydrological Applications - 2nd Edition
Hans van der Kwast & Kurt Menke

Introduction to QGIS
Scott Madry Ph.D.

Leaflet Cookbook
Numa Gremling

QGIS Map Design - 2nd Edition
Anita Graser & Gretchen N. Peterson

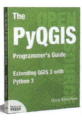

The PyQGIS Programmer's Guide 3
Gary Sherman

pgRouting
Regina O. Obe & Leo S. Hsu

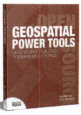

Geospatial Power Tools
Tyler Mitchell, GDAL Developers

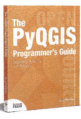

The PyQGIS Programmer's Guide
Gary Sherman

The Geospatial Desktop
Gary Sherman

locatepress.com

OPEN SOURCE
GEOSPATIAL | BOOKS

📞 +1 (250) 303-1831   ✉ tyler@locatepress.com